The Fairyland of Geometry

Simon Newcomb, 1905

THE FAIRYLAND OF GEOMETRY
and Other Essays in Science

Simon Newcomb

Edited, and with Notes and Commentary, by
David Stover

Milestones in Science and Discovery
Rock's Mills Press

Frontispiece: *Photograph of Simon Newcomb, 1905,*
Harris & Ewing Collection, Library of Congress, Washington, DC.
Cover Image: *www.freeimages.co.uk*

PUBLISHED BY
Rock's Mills Press

Copyright © 2016 by David Stover
Library and Archives Canada Cataloguing in Publication data is available from the publisher. Contact us at customer.service@rocksmillspress.com.

www.rocksmillspress.com

Contents

PREFACE *by David Stover*   *vii*

1   A View of the Universe   *1*
2   What the Astronomers Are Doing   *6*
3   The New Problems of the Universe   *16*
4   Life in the Universe   *21*
5   Constellation and Star Names   *30*
6   The Universe as an Organism   *36*
7   The Coming Total Eclipse of the Sun   *44*
8   How the Planets Are Weighed   *58*
9   The Sun   *63*
10   The Earth   *73*
11   The Moon   *81*
12   The Planet Mars   *89*
13   The Fairyland of Geometry   *98*
14   The Relation of Scientific Method to Social Progress   *105*
15   Is the Airship Coming?   *117*
16   Modern Occultism   *125*
17   Can We Make It Rain?   *141*
18   What Is a Liberal Education?   *147*

EPILOGUE: What Happened Next   *151*
AFTERWORD: Simon Newcomb, Scientist and Popularizer   *161*
EDITOR'S NOTES   *165*

Preface

Simon Newcomb rose from poverty to become one of the most distinguished astronomers of the latter half of the nineteenth century. Born in Nova Scotia, he fled to the United States to escape an oppressive apprenticeship. Once there, he earned a degree from Harvard, worked for the branch of the U.S. Navy that produced the nautical almanac—that era's essential tool for navigation at sea—and in 1861 was appointed professor of mathematics at the U.S. Naval Observatory in Washington, where he would remain for more than 35 years. His work on calculating tables of lunar and planetary motions and putting them on a sounder footing continued until just before his death in 1909, and is considered his greatest contribution to astronomy, though he was involved in a good deal of other work, both scientific and administrative, during his long career.

What made Newcomb unusual was the fact that he was not only a distinguished scientist but a well-known one—indeed, perhaps the best-known American scientist of the age, and one of the era's preeminent popularizers of science. A prolific writer, he produced not only dozens of research papers but hundreds of popular works for newspapers, magazines, encyclopedias, and book publishers. Most of them dealt with astronomical and mathematical subjects, but he was not afraid to venture further afield, into economics, education, politics, technology, even occultism. His *Popular Astronomy* (1878) became a standard text and *Astronomy for Everybody* (1902) sold tens of thousands of copies and remained in print long after his death.

My main criterion for picking the essays that are included in this collection was that they still have something interesting and useful to say to the modern reader. In a few cases, the interest lies in the contrast between what Newcomb thought to be true and what we know today. His famous essay on the impossibility of heavier-than-air flight falls into this category and remains a compelling example of the pitfalls of technological prognostication. The discussions of the state of astronomy at the dawn of the twentieth century shed light on how far knowledge has advanced in little more than a lifetime. But most of the pieces that follow are still worth reading on their own terms. Newcomb's description of a solar eclipse is as evocative as any written since. The debate about educational priorities rages as fiercely in our day as in his, and he still has something important to say on the matter. A vast amount of work has been done in non-Euclidean geometry since he wrote "The Fairyland of Geometry"—this collection's title piece—yet his overview of the subject's basics remains as clear as when it was first published in 1902. And his reasoned skepticism about the claims of "modern occultism" (we would call it "parapsychology") is as refreshing now as it was a century ago.

The enduring value of Newcomb's science popularizations lies not in what they tell us about the universe as such—there are many more up-to-date places to find *that*—but in the insight they provide into how that knowledge was arrived at, and how each generation makes sense of the world in its own unique way. And there is always pleasure to be found in a first-rate mind explaining difficult ideas clearly and compellingly. In doing so, a skilled science writer—and Newcomb was one of the best—evokes in his readers the same thrill at learning something new that is central to science itself. The frontiers of science may have moved on since Newcomb set down his pen for the last time, but the spirit of discovery embodied in these essays remains fresh.

—David Stover

The Fairyland of Geometry

CHAPTER ONE

A View of the Universe

"A View of the Universe" is the opening chapter in Newcomb's 1902 book Astronomy for Everybody: A Popular Exposition of the Wonders of the Heavens, *which was based in part on articles previously published in* McClure's Magazine *and other outlets.* Astronomy for Everybody *was Newcomb's best-selling book and remained in print for the better part of four decades; later editions were revised by Robert H. Baker of the University of Illinois Observatory. When Newcomb first wrote this essay, the general assumption was that what we now call the Milky Way galaxy—the giant grouping of stars to which our sun belongs—was all there was to the universe. Not till the 1920s was the existence of other galaxies beyond the Milky Way established on a firm footing, though it had long been suspected by at least some astronomers. (Indeed, Newcomb himself alludes to the possibility early in this piece.) That being said, if you substitute the word "galaxy" for "universe" in the chapter's title and first sentence, what follows is as evocative a portrayal of the minuteness of humanity's domain and the vastness of the starry universe as anything written since. And with the twenty-first century demotion of Pluto (discovered 30 years after this chapter was written) from planetary status, even Newcomb's reference to the eight planets of the solar system is once again correct—though of course the solar system also includes a multitude of smaller bodies utterly unknown in Newcomb's day.*

L et us enter upon our subject by taking a general view of this universe in which we live, fancying ourselves looking at it from a point without its limits. Far away, indeed, is the point we must choose. To give a conception of the distance, let us measure it by the motion of light. This agent, darting through 186,000 miles in every second, would make the circuit of the earth several times between two ticks of a watch. The standpoint which we choose will probably be well situated if we take it at a distance through which light would travel in 100,000 years. So far as we know, we should at this point find ourselves in utter darkness, a black and starless sky surrounding us on all sides. But, in one direction, we should see a large patch of feeble light spreading over a considerable part of the heavens like a faint cloud or the first glimmer of a dawn. Possibly there might be other such patches in different directions, but of these we know nothing. The one which we have mentioned, and which we call the universe, is that which we are to inspect. We therefore fly toward it—how fast we need not say. To reach it in a month we should have to go a million times as fast as light. As we approach, it continually spreads out over more of the black sky which it at length half covers, the region behind us being still entirely black.

Before reaching this stage we begin to see points of light glimmering here and there in the mass. Continuing our course, these points become more numerous, and seem to move past us and disappear behind us in the distance, while new ones continually come into view in front, as the passengers on a railway train see landscape and houses flit by them. These are stars, which, when we get well in among them, stud the whole heavens as we see them do at night. We might pass through the whole cloud at the enormous speed we have fancied, without seeing anything but stars and, perhaps, a few great nebulous masses of foggy light scattered here and there among them.

But instead of doing this, let us select one particular star and slacken our speed to make a closer inspection of it. This one is rather a small star; but as we approach it, it seems to our eyes to grow brighter. In time it shines like Venus. Then it casts a shadow; then we can read by its light; then it begins to dazzle our eyes. It looks like a little sun. It is the Sun!

Let us get into a position which, compared with the distances we have been travelling, is right alongside of the sun, though, expressed in our ordinary measure, it may be a thousand million miles away. Now, looking down and around us, we see eight star-like points scattered around the sun at different distances. If we watch them long enough we shall see them all in motion around the Sun, completing their circuit in times ranging from three months to more than 160 years. They move at very different distances; the most distant is seventy times as far as the nearest.

These star-like bodies are the planets. By careful examination we see that they differ from the stars in being opaque bodies, shining only by light borrowed from the sun.

Let us pay one of them a visit. We select the third in order from the sun. Approaching it in a direction which we may call from above, that is to say from a direction at right angles to the line drawn from it to the sun, we see it grow larger and brighter as we get nearer. When we get very near, we see it looking like a half-moon—one hemisphere being in darkness and the other illuminated by the sun's rays. As we approach yet nearer, the illuminated part, always growing larger to our sight, assumes a mottled appearance. Still expanding, this appearance gradually resolves itself into oceans and continents, obscured over perhaps half their surface by clouds. The surface upon which we are looking continually spreads out before us, filling more and more of the sky, until we see it to be a world. We land upon it, and here we are upon the earth.

Thus, a point which was absolutely invisible while we were flying through the celestial spaces, which became a star when we got near the sun, and an opaque globe when yet nearer, now becomes the world on which we live.

This imaginary flight makes known to us a capital fact of astronomy: The great mass of stars which stud the heavens at night are suns. To express the idea in another way, the sun is merely one of the stars. Compared with its fellows it is rather a small one, for we know of stars that emit thousands or even tens of thousands of times the light and heat of the sun. Measuring things simply by their intrinsic importance, there is nothing special to distinguish our sun from the hundreds of millions of its companions. Its importance to us and its comparative greatness in our eyes arise simply from the accident of our relation to it.

The great universe of stars which we have described looks to us from the earth just as it looked to us during our imaginary flight through it. The stars which stud our sky are the same stars which we saw on our flight. The great difference between our view of the heavens and the view from a point in the starry distances is the prominent position occupied by the sun and planets. The former is so bright that during the daytime it completely obliterates the stars. If we could cut off the sun's rays from any very wide region, we should see the stars around the sun in the daytime as well as by night. These bodies surround us in all directions as if the earth were placed in the centre of the universe, as was supposed by the ancients.

We may connect what we have just learned about the universe at large with what we see in the heavens. What we call the heavenly bodies are of two classes. One of these comprises the millions of stars the arrangement and appearance of which we have just described. The other comprises a single star, which is for us the most important of all, and the bodies connected with it. This collection of bodies, with the sun in its center, forms a little colony all by itself, which we call the solar system. The feature of this system which I wish first to impress on the reader's mind is its very small dimensions when compared with the distances between the stars. All around it are spaces which, so far as we yet know, are quite void through enormous distances. If we could fly across the whole breadth of the system, we should not be able to see that we were any nearer the stars in front of us, nor would the constellations look in any way different from what they do from our earth. An astronomer armed with the finest instruments would be able to detect a change only by the most exact observations, and then only in the case of the nearer stars.

A conception of the respective magnitudes and distances of the heavenly bodies, which will help the reader in conceiving of the universe as it is, may be gained by supposing us to look at a little model of it. Let us imagine that, in this model of the universe, the earth on which we dwell is represented by a grain of mustard seed. The moon will then be a particle about one-fourth the diameter of the grain, placed at a distance of an inch from the earth. The sun will be represented by a large apple, placed at a distance of forty feet. Other planets, ranging in size from an invisible particle to a pea, must be imagined at distances from the sun varying from ten feet to a quarter of a mile. We must then imagine all these little objects to be slowly moving around the sun at their respective distances, in times varying from three months to 160 years. As the mustard seed performs its revolution in the course of a year we must imagine the moon to accompany it, making a revolution around it every month.

On this scale a plan of the whole solar system can be laid down in a field half-a-mile square. Outside of this field we should find a tract broader than the whole continent of America without a visible object in it unless perhaps comets scattered around its border. Far beyond the limits of the American continent we should find the nearest star, which, like our sun, might be represented by a large apple. At still greater distances, in every direction, would be other stars, but, in the general average, they would be separated from each other as widely as the nearest star is from the sun. A region of the little model as large as the whole earth might contain only two or three stars.

We see from this how, in a flight through the universe, like the one we have imagined, we might overlook such an insignificant little body as our earth, even if we made a careful search for it. We should be like a person flying through the Mississippi Valley, looking for a grain of mustard seed which he knew was hidden somewhere on the American continent. Even the bright shining apple representing the sun might be overlooked unless we happened to pass quite near it.

CHAPTER TWO

What the Astronomers Are Doing

In this article, originally published in Harper's Magazine *in July 1902 and then reprinted in Newcomb's essay collection* Side-lights on Astronomy *(1906), Newcomb presents a concise, readable overview of the state of astronomical knowledge at the dawn of the twentieth century. Human beings like to divide history into convenient and comforting periods, but history itself is a continuous process and, with the benefit of hindsight, a snapshot taken at one particular point reveals fascinating hints of things to come. A number of the topics Newcomb touches on here—the nature of the sun and the wellsprings of its energy, or the overall structure of our universe—turned out to be major areas of investigation during the course of the twentieth century. And while the intricacies of stellar evolution were unknown in Newcomb's day, his brief, broad-brushstroke description of a star's life story on page 12 is still essentially correct. Stars begin as vast, diffuse clouds of gas and end as cold, dark, dense cinders. But what is perhaps most noteworthy is the emphasis in nineteenth- and early twentieth-century astronomy on mapping the heavens. In the hundred years that followed, the focus shifted from the process of map-creation (though, of course, we possess today maps of the cosmos of an accuracy that would have staggered astronomers of a hundred years ago) to achieving a deeper understanding of what the items marked on the map actually were, and how they came to be.*

I n no field of science has human knowledge been more extended in our time than in that of astronomy. Forty years ago [i.e., about 1860] astronomical research seemed quite barren of results of great interest or value to our race. The observers of the world were working on a traditional system, grinding out results in an endless course, without seeing any prospect of the great generalizations to which they might ultimately lead. Now this is all changed. A new instrument, the spectroscope, has been developed, the extent of whose revelations we are just beginning to learn, although it has been more than thirty years in use. The application of photography has been so extended that, in some important branches of astronomical work, the observer simply photographs the phenomenon which he is to study, and then makes his observation on the developed negative.

The world of astronomy is one of the busiest that can be found today, and the writer proposes, with the reader's courteous consent, to take him on a stroll through it and see what is going on. We may begin our inspection with a body which is, for us, next to the earth, the most important in the universe. I mean the sun. At the Greenwich Observatory the sun has for more than twenty years been regularly photographed on every clear day, with the view of determining the changes going on in its spots. In recent years these observations have been supplemented by others, made at stations in India and Mauritius, so that by the combination of all it is quite exceptional to have an entire day pass without at least one photograph being taken. On these observations must mainly rest our knowledge of the curious cycle of change in the solar spots, which goes through a period of about eleven years, but of which no one has as yet been able to establish the cause.

This Greenwich system has been extended and improved by an American. Professor George E. Hale, formerly Director of the Yerkes Observatory, has devised an instrument for taking photographs of the sun by a single ray of the spectrum. The light emitted by calcium, the base of lime, and one of the substances most abundant in the sun, is often selected to impress the plate.

The Carnegie Institution has recently organized an enterprise for carrying on the study of the sun under a combination of better conditions than were ever before enjoyed. The first requirement in such a case is the ablest and most enthusiastic worker in the field, ready to devote all his energies to its cultivation. This requirement is found in the person of Professor Hale himself. The next requirement is an atmosphere of the greatest transparency, and a situation at a high elevation above sea-level, so that the passage of light from the sun to the observer shall be obstructed as little as possible by the mists and vapors near the earth's surface. This requirement is reached by placing the observatory on Mount Wilson, near Pasadena, California, where the climate is found to be the best of any in the United States, and probably not exceeded by that of any other attainable point in the world. The third requirement is the best of instruments, specially devised to meet the requirements. In this respect we may be sure that nothing attainable by human ingenuity will be found wanting.

Thus provided, Professor Hale has entered upon the task of studying the sun, and recording from day to day all the changes going on in it, using specially devised instruments for each purpose in view. Photography is made use of through almost the entire investigation. A full description of the work would require an enumeration of technical details, into which we need not enter at present. Let it, therefore, suffice to say in a general way that the study of the sun is being carried on on a scale and with an energy worthy of the most important subject that presents itself to the astronomer. Closely associated with this work is that of Professor Langley and Dr. Abbot, at the Astro-Physical Observatory of the Smithsonian Institution, who have recently completed one of the most important works ever carried out on the light of the sun. They have for years been analyzing those of its rays which, although entirely invisible to our eyes, are of the same nature as those of light, and are felt by us as heat. To do this, Langley invented a sort of artificial eye, which he called a bolometer, in which the optic nerve is made of an extremely thin strip of metal, so slight that one can hardly see it, which is traversed by an electric current. This eye would be so dazzled by the heat radiated from one's body that, when in use, it must be protected from all such heat by being enclosed in a case kept at a constant temperature by being immersed in water. With this eye the two observers have mapped the heat rays of the sun down to an extent and with a precision which were before entirely unknown.

The question of possible changes in the sun's radiation, and of the relation of those changes to human welfare, still eludes our scrutiny. With all the efforts that have been made, the physicist of today has not yet been able to make anything like an exact determination of the total amount of heat received from the sun. The largest measurements are almost double the smallest. This is partly due to the atmosphere absorbing an unknown and variable fraction of the sun's rays which pass through it, and partly to the difficulty of distinguishing the heat radiated by the sun from that radiated by terrestrial objects.

In one recent instance, a change in the sun's radiation has been noticed in various parts of the world, and is of especial interest because there seems to be little doubt as to its origin. In the latter part of 1902 an extraordinary diminution was found in the intensity of the sun's heat, as measured by the bolometer and other instruments. This continued through the first part of 1903, with wide variations at different places, and it was more than a year after the first diminution before the sun's rays again assumed their ordinary intensity.

This result is now attributed to the eruption of Mount Pélée, during which an enormous mass of volcanic dust and vapor was projected into the higher regions of the air, and gradually carried over the entire earth by winds and currents. Many of our readers may remember that something yet more striking occurred after the great cataclysm at Krakatoa in 1883, when, for more than a year, red sunsets and red twilights of a depth of shade never before observed were seen in every part of the world.

What we call universology—the knowledge of the structure and extent of the universe—must begin with a study of the starry heavens as we see them. There are perhaps one hundred million stars in the sky within the reach of telescopic vision. This number is too great to allow of all the stars being studied individually; yet, to form the basis for any conclusion, we must know the positions and arrangement of as many of them as we can determine.

To do this the first want is a catalogue giving very precise positions of as many of the brighter stars as possible. The principal national observatories, as well as some others, are engaged in supplying this want. Up to the present time about 200,000 stars visible in our latitudes have been catalogued on this precise plan, and the work is still going on. In that part of the sky which we never see, because it is only visible from the southern hemisphere, the corresponding work is far from being as extensive. Sir David Gill, astronomer at the Cape of Good Hope, and also the directors of other southern observatories, are engaged in pushing it forward as rapidly as the limited facilities at their disposal will allow.

Next in order comes the work of simply listing as many stars as possible. Here the most exact positions are not required. It is only necessary to lay down the position of each star with sufficient exactness to distinguish it from all its neighbors. About 400,000 stars were during the last half-century listed in this way at the observatory of Bonn by Argelander, Schönfeld, and their assistants. This work is now being carried through the southern hemisphere on a large scale by Thome, Director of the Cordoba Observatory, in the Argentine Republic. This was founded thirty years ago by our Dr. B. A. Gould, who turned it over to Dr. Thome in 1886. The latter has, up to the present time, fixed and published the positions of nearly half a million stars. This work of Thome extends to fainter stars than any other yet attempted, so that, as it goes on, we have more stars listed in a region invisible in middle northern latitudes than we have for that part of the sky we can see. Up to the present time three quarto volumes giving the positions and magnitudes of the stars have appeared. Two or three volumes more, and, perhaps, ten or fifteen years, will be required to complete the work.

About twenty years ago it was discovered that, by means of a telescope especially adapted to this purpose, it was possible to photograph many more stars than an instrument of the same size would show to the eye. This discovery was soon applied in various quarters. Sir David Gill, with characteristic energy, photographed the stars of the southern sky to the number of nearly half a million. As it was beyond his power to measure off and compute the positions of the stars from his plates, the latter were sent to Professor J. C. Kapteyn, of Holland, who undertook the enormous labor of collecting them into a catalogue, the last volume of which was published in 1899. One curious result of this enterprise is that the work of listing the stars is more complete for the southern hemisphere than for the northern.

Another great photographic work now in progress has to do with the millions of stars which it is impossible to handle individually. Fifteen years ago an association of observatories in both hemispheres undertook to make a photographic chart of the sky on the largest scale. Some portions of this work are now approaching completion, but in others it is still in a backward state, owing to the failure of several South American observatories to carry out their part of the programme. When it is all done we shall have a picture of the sky, the study of which may require the labor of a whole generation of astronomers.

Quite independently of this work, the Harvard University, under the direction of Professor Pickering, keeps up the work of photographing the sky on a surprising scale. On this plan we do not have to leave it to posterity to learn whether there is any change in the heavens, for one result of the enterprise has been the discovery of thirteen of the new stars which now and then blaze out in the heavens at points where none were before known. Professor Pickering's work has been continually enlarged and improved until about 150,000 photographic plates, showing from time to time the places of countless millions of stars among their fellows, are now stored at the Harvard Observatory. Not less remarkable than this wealth of material has been the development of skill in working it up. Some idea of the work will be obtained by reflecting that, thirty years ago, careful study of the heavens by astronomers devoting their lives to the task had resulted in the discovery of some two or three hundred stars varying in their light. Now, at Harvard, through keen eyes studying and comparing successive photographs not only of isolated stars, but of clusters and agglomerations of stars in the Milky Way and elsewhere, discoveries of such objects numbering hundreds have been made, and the work is going on with ever-increasing speed. Indeed, the number of variable stars now known is such that their study as individual objects no longer suffices, and they must hereafter be treated statistically with reference to their distribution in space, and their relations to one another, as a census classifies the entire population without taking any account of individuals.

The works just mentioned are concerned with the stars. But the heavenly spaces contain nebulae as well as stars; and photography can now be even more successful in picturing them than the stars. A few years ago the late lamented Keeler, at the Lick Observatory, undertook to see what could be done by pointing the Crossley reflecting telescope at the sky and putting a sensitive photographic plate in the focus. He was surprised to find that a great number of nebulae, the existence of which had never before been suspected, were impressed on the plate. Up to the present time the positions of about 8000 of these objects have been listed. Keeler found that there were probably 200,000 nebulae in the heavens capable of being photographed with the Crossley reflector. But the work of taking these photographs is so great, and the number of reflecting telescopes which can be applied to it so small, that no one has ventured to seriously commence it. It is worthy of remark that only a very small fraction of these objects which can be photographed are visible to the eye, even with the most powerful telescope.

This demonstration of what the reflecting telescope can do may be regarded as one of the most important discoveries of our time as to the capabilities of astronomical instruments. It has long been known that the image formed in the focus of the best refracting telescope is affected by an imperfection arising from the different action of the glasses on rays of light of different colors. Hence, the image of a star can never be seen or photographed with such an instrument, as an actual point, but only as a small, diffused mass. This difficulty is avoided in the reflecting telescope; but a new difficulty is found in the bending of the mirror under the influence of its own weight. Devices for overcoming this had been so far from successful that, when Mr. Crossley presented his instrument to the Lick Observatory, it was feared that little of importance could be done with it. But as often happens in human affairs outside the field of astronomy, when ingenious and able men devote their attention to the careful study of a problem, it was found that new results could be reached. Thus it was that, before a great while, what was supposed to be an inferior instrument proved not only to have qualities not before suspected, but to be the means of making an important addition to the methods of astronomical investigation.

In order that our knowledge of the position of a star may be complete, we must know its distance. This can be measured only through the star's parallax—that is to say, the slight change in its direction produced by the swing of our earth around its orbit. But so vast is the distance in question that this change is immeasurably small, except for, perhaps, a few hundred stars, and even for these few its measurement almost baffles the skill of the most expert astronomer. Progress in this direction is therefore very slow, and there are probably not yet a hundred stars of which the parallax has been ascertained with any approach to certainty. Dr. Chase is now completing an important work of this kind at the Yale Observatory.

To the most refined telescopic observations, as well as to the naked eye, the stars seem all alike, except that they differ greatly in brightness, and somewhat in color. But when their light is analyzed by the spectroscope, it is found that scarcely any two are exactly alike. An important part of the work of the astrophysical observatories, especially that of Harvard, consists in photographing the spectra of thousands of stars, and studying the peculiarities thus brought out. At Harvard a large portion of this work is done as part of the work of the Henry Draper Memorial, established by his widow in memory of the eminent investigator of New York, who died twenty years ago.

By a comparison of the spectra of stars Sir William Huggins has developed the idea that these bodies, like human beings, have a life history. They are nebulae in infancy, while the progress to old age is marked by a constant increase in the density of their substance. Their temperature also changes in a way analogous to the vigor of the human being. During a certain time the star continually grows hotter and hotter. But an end to this must come, and it cools off in old age. What the age of a star may be is hard even to guess. It is many millions of years, perhaps hundreds, possibly even thousands, of millions.

Some attempt at giving the magnitude is included in every considerable list of stars. The work of determining the magnitudes with the greatest precision is so laborious that it must go on rather slowly. It is being pursued on a large scale at the Harvard Observatory, as well as in that of Potsdam, Germany.

We come now to the question of changes in the appearance of bright stars. It seems pretty certain that more than one per cent of these bodies fluctuate to a greater or less extent in their light. Observations of these fluctuations, in the case of at least the brighter stars, may be carried on without any instrument more expensive than a good opera-glass—in fact, in the case of stars visible to the naked eye, with no instrument at all.

As a general rule, the light of these stars goes through its changes in a regular period, which is sometimes as short as a few hours, but generally several days, frequently a large fraction of a year or even eighteen months. Observations of these stars are made to determine the length of the period and the law of variation of the brightness. Any person with a good eye and skill in making estimates can make the observations if he will devote sufficient pains to training himself; but they require a degree of care and assiduity which is not to be expected of any one but an enthusiast on the subject. One of the most successful observers of the present time is Mr. W. A. Roberts, a resident of South Africa, whom the Boer war did not prevent from keeping up a watch of the southern sky, which has resulted in greatly increasing our knowledge of variable stars. There are also quite a number of astronomers in Europe and America who make this particular study their specialty.

During the past fifteen years the art of measuring the speed with which a star is approaching us or receding from us has been brought to a wonderful degree of perfection. The instrument with which this was first done was the spectroscope; it is now replaced with another of the same general kind, called the spectrograph. The latter differs from the other only in that the spectrum of the star is photographed, and the observer makes his measures on the negative. This method was first extensively applied at the Potsdam Observatory in Germany, and has lately become one of the specialties of the Lick Observatory, where Professor Campbell has brought it to its present degree of perfection. The Yerkes Observatory is also beginning work in the same line, where Professor Frost is already rivalling the Lick Observatory in the precision of his measures.

Let us now go back to our own little colony and see what is being done to advance our knowledge of the solar system. This consists of planets, on one of which we dwell, moons revolving around them, comets, and meteoric bodies. The principal national observatories keep up a more or less orderly system of observations of the positions of the planets and their satellites in order to determine the laws of their motion. As in the case of the stars, it is necessary to continue these observations through long periods of time in order that everything possible to learn may be discovered.

Our own moon is one of the enigmas of the mathematical astronomer. Observations show that she is deviating from her predicted place, and that this deviation continues to increase. True, it is not very great when measured by an ordinary standard. The time at which the moon's shadow passed a given point near Norfolk during the total eclipse of May 29, 1900, was only about seven seconds different from the time given in the *Astronomical Ephemeris*. The path of the shadow along the earth was not out of place by more than one or two miles. But, small though these deviations are, they show that something is wrong, and no one has as yet found out what it is. Worse yet, the deviation is increasing rapidly. The observers of the total eclipse in August, 1905, were surprised to find that it began twenty seconds before the predicted time. The mathematical problems involved in correcting this error are of such complexity that it is only now and then that a mathematician turns up anywhere in the world who is both able and bold enough to attack them.

There now seems little doubt that Jupiter is a miniature sun, only not hot enough at its surface to shine by its own light. The point in which it most resembles the sun is that its equatorial regions rotate in less time than do the regions near the poles. This shows that what we see is not a solid body. But none of the careful observers have yet succeeded in determining the law of this difference of rotation.

Twelve years ago a suspicion which had long been entertained that the earth's axis of rotation varied a little from time to time was verified by Chandler. The result of this is a slight change in the latitude of all places on the earth's surface, which admits of being determined by precise observations. The National Geodetic Association has established four observatories on the same parallel of latitude—one at Gaithersburg, Maryland, another on the Pacific coast, a third in Japan, and a fourth in Italy—to study these variations by continuous observations from night to night. This work is now going forward on a well-devised plan.

A fact which will appeal to our readers on this side of the Atlantic is the success of American astronomers. Sixty years ago it could not be said that there was a well-known observatory on the American continent. The cultivation of astronomy was confined to a professor here and there, who seldom had anything better than a little telescope with which he showed the heavenly bodies to his students. But during the past thirty years all this has been changed. The total quantity of published research is still less among us than on the continent of Europe, but the number of men who have reached the highest success among us may be judged by one fact. The Royal Astronomical Society of England awards an annual medal to the English or foreign astronomer deemed most worthy of it. The number of these medals awarded to Americans within twenty-five years is about equal to the number awarded to the astronomers of all other nations foreign to the English. That this preponderance is not growing less is shown by the award of medals to Americans in three consecutive years—1904, 1905, and 1906. The recipients were Hale, Boss, and Campbell. Of the fifty foreign associates chosen by this society for their eminence in astronomical research, no less than eighteen—more than one-third—are Americans.

CHAPTER THREE

The New Problems of the Universe

If in the previous chapter Newcomb looked back over nineteenth-century developments in astronomy, here he looks forward to puzzles that had been left for the new century to solve. "The New Problems of the Universe" was published in Harper's *in November 1903 and reprinted in* Side-lights on Astronomy. *Newcomb's crystal ball was sometimes clear, sometimes cloudy, and this essay is no exception. He shows considerable insight in declaring that the final task of the old century seemed to be "setting forth more [scientific] problems for this century to solve than it has ever itself succeeded in mastering." But the suggestion that the twentieth century would likely not produce anything as important in the realm of invention as the steamship or the telegraph was rather less prescient! The version of the essay reprinted here has been edited to remove a long and, for the modern reader, not overly interesting discussion of novae, or new stars. The effect of the excision is to sharpen the focus on Newcomb's most important prediction about astronomical research in the century to come: that some of its most profound findings would come through the study of forces and effects hitherto unknown.*

The achievements of the nineteenth century are still a theme of congratulation on the part of all who compare the present state of the world with that of one hundred years ago. And yet, if we should fancy the most sagacious prophet, endowed with a brilliant imagination, to have set forth in the year 1806 the problems that the century might solve and the things which it might do, we should be surprised to see how few of his predictions had come to pass. He might have fancied aerial navigation and a number of other triumphs of the same class, but he would hardly have had either steam navigation or the telegraph in his picture. In 1856 an article appeared in *Harper's Magazine* depicting some anticipated features of life in A.D. 3000. We have since made great advances, but they bear little resemblance to what the writer imagined. He did not dream of the telephone, but did describe much that has not yet come to pass and probably never will.

The fact is that, much as the nineteenth century has done, its last work was to amuse itself by setting forth more problems for this century to solve than it has ever itself succeeded in mastering. We should not be far wrong in saying that today there are more riddles in the universe than there were before men knew that it contained anything more than the objects they could see.

So far as mere material progress is concerned, it may be doubtful whether anything so epoch-making as the steam engine or the telegraph is held in store for us by the future. But in the field of purely scientific discovery we are finding a crowd of things of which our philosophy did not dream even ten years ago.

The greatest riddles which the nineteenth century has bequeathed to us relate to subjects so widely separated as the structure of the universe and the structure of atoms of matter. We see more and more of these structures, and we see more and more of unity everywhere, and yet new facts difficult of explanation are being added more rapidly than old facts are being explained.

We all know that the nineteenth century was marked by a separation of the sciences into a vast number of specialties, to the subdivisions of which one could see no end. But the great work of the twentieth century will be to combine many of these specialties. The physical philosopher of the present time is directing his thought to the demonstration of the unity of creation. Astronomical and physical researches are now being united in a way which is bringing the infinitely great and the infinitely small into one field of knowledge. Ten years ago the atoms of matter, of which it takes millions of millions to make a drop of water, were the minutest objects with which science could imagine itself to be concerned, Now a body of experimentalists, prominent among whom stand Professors J. J. Thompson, Becquerel, and Röntgen, have demonstrated the existence of objects so minute that they find their way among and between the atoms of matter as raindrops do among the buildings of a city. More wonderful yet, it seems likely, although it has not been demonstrated, that these little things, called "corpuscles," play an important part in what is going on among the stars. Whether this be true or not, it is certain that there do exist in the universe emanations of some sort, producing visible effects, the investigation of which the nineteenth century has had to bequeath to the twentieth.

For the purpose of the navigator, the direction of the magnetic needle is invariable in any one place, for months and even years; but when exact scientific observations on it are made, it is found subject to numerous slight changes. The most regular of these consists in a daily change of its direction. It moves one way from morning until noon, and then, late in the afternoon and during the night, turns back again to its original pointing. The laws of this change have been carefully studied from observations, which show that it is least at the equator and larger as we go north into middle latitudes; but no explanation of it resting on an indisputable basis has ever been offered.

Besides these regular changes, there are others of a very irregular character. Every now and then the changes in the direction of the magnet are wider and more rapid than those which occur regularly every day. The needle may move back and forth in a way so fitful as to show the action of some unusual exciting cause. Such movements of the needle are commonly seen when there is a brilliant aurora. This connection shows that a magnetic storm and an aurora must be due to the same or some connected causes.

Those of us who are acquainted with astronomical matters know that the number of spots on the sun goes through a regular cycle of change, having a period of eleven years and one or two months. Now, the curious fact is, when the number and violence of magnetic storms are recorded and compared, it is found that they correspond to the spots on the sun, and go through the same period of eleven years. The conclusion seems almost inevitable: magnetic storms are due to some emanation sent out by the sun, which arises from the same cause that produces the spots. . . .

But when we ask by what agency it is possible for the sun to affect the magnetism of the earth, and when we trace the passage of some agent between the two bodies, we find nothing to explain the action. To all appearance, the space between the earth and the sun is a perfect void. That electricity cannot of itself pass through a vacuum seems to be a well-established law of physics. It is true that electromagnetic waves, which are supposed to be of the same nature with those of light, and which are used in wireless telegraphy, do pass through a vacuum and may pass from the sun to the earth. But there is no way of explaining how such waves would either produce or affect the magnetism of the earth.

The mysterious emanations from various substances, under certain conditions, may have an intimate relation with yet another of the mysteries of the universe. It is a fundamental law of the universe that when a body emits light or heat, or anything capable of being transformed into light or heat, it can do so only by the expenditure of force, limited in supply. The sun and stars are continually sending out a flood of heat. They are exhausting the internal supply of something which must be limited in extent. Whence comes the supply? How is the heat of the sun kept up? If it were a hot body cooling off, a very few years would suffice for it to cool off so far that its surface would become solid and very soon cold. In recent years, the theory universally accepted has been that the supply of heat is kept up by the continual contraction of the sun, by mutual gravitation of its parts as it cools off. This theory has the advantage of enabling us to calculate, with some approximation to exactness, at what rate the sun must be contracting in order to keep up the supply of heat which it radiates. On this theory, it must, ten million years ago, have had twice its present diameter, while less than twenty million years ago it could not have existed except as an immense nebula filling the whole solar system. We must bear in mind that this theory is the only one which accounts for the supply of heat, even through human history. If it be true, then the sun, earth, and solar system must be less than twenty million years old.

Here the geologists step in and tell us that this conclusion is wholly inadmissible. The study of the strata of the earth and of many other geological phenomena, they assure us, makes it certain that the earth must have existed much in its present condition for hundreds of millions of years. During all that time there can have been no great diminution in the supply of heat radiated by the sun.

The astronomer, in considering this argument, has to admit that he finds a similar difficulty in connection with the stars and nebulas. It is an impossibility to regard these objects as new; they must be as old as the universe itself. They radiate heat and light year after year. In all probability, they must have been doing so for millions of years. Whence comes the supply? The geologist may well claim that until the astronomer explains this mystery in his own domain, he cannot declare the conclusions of geology as to the age of the earth to be wholly inadmissible.

Now, the scientific experiments of the last two years have brought this mystery of the celestial spaces right down into our earthly laboratories. Monsieur and Madame Curie have discovered the singular metal radium, which seems to send out light, heat, and other rays incessantly, without, so far as has yet been determined, drawing the required energy from any outward source. As we have already pointed out, such an emanation must come from some storehouse of energy. Is the storehouse, then, in the medium itself, or does the latter draw it from surrounding objects? If it does, it must abstract heat from these objects. This question has been settled by Professor Dewar, at the Royal Institution, London, by placing the radium in a medium next to the coldest that art has yet produced— liquid air. The latter is surrounded by the only yet colder medium, liquid hydrogen, so that no heat can reach it. Under these circumstances, the radium still gives out heat, boiling away the liquid air until the latter has entirely disappeared. Instead of the radiation diminishing with time, it rather seems to increase.

Called on to explain all this, science can only say that a molecular change must be going on in the radium, to correspond to the heat it gives out. What that change may be is still a complete mystery. It is a mystery which we find alike in those minute specimens of the rarest of substances under our microscopes, in the sun, and in the vast nebulous masses in the midst of which our whole solar system would be but a speck. The unravelling of this mystery must be the great work of science of the twentieth century. What results shall follow for mankind one cannot say, any more than he could have said two hundred years ago what modern science would bring forth. Perhaps, before future developments, all the boasted achievements of the nineteenth century may take the modest place which we now assign to the science of the eighteenth century—that of the infant which is to grow into a man.

CHAPTER FOUR

Life in the Universe

*Exobiology—the study of life elsewhere in the universe—has been de-
scribed as a science without a subject, considering the fact that, so far,
we haven't come across life anywhere other than on the earth. In that
respect nothing has changed since Newcomb wrote this essay, which was
published in* Harper's *in August 1905. Of course we know far more today
about the nature of life, and also about the environmental conditions that
exist on the other planets and satellites of the solar system (and even,
over the past two decades, the likely environmental conditions on planets
of other solar systems). But Newcomb's conclusions still ring true: that
life is adaptable to an amazingly wide range of environments; and that
given the immensity of the universe, the odds are good that not only living organisms but perhaps even intelligent beings exist elsewhere in
space.*

So far as we can judge from what we see on our globe, the production
of life is one of the greatest and most incessant purposes of nature.
Life is absent only in regions of perpetual frost, where it never has
an opportunity to begin; in places where the temperature is near the boil-
ing-point, which is found to be destructive to it; and beneath the earth's
surface, where none of the changes essential to it can come about. Within
the limits imposed by these prohibitory conditions—that is to say, within
the range of temperature at which water retains its liquid state, and in regions where the sun's rays can penetrate and where wind can blow and
water exist in a liquid form—life is the universal rule. How prodigal nature
seems to be in its production is too trite a fact to be dwelt upon. We have

all read of the millions of germs which are destroyed for every one that comes to maturity. Even the higher forms of life are found almost everywhere. Only small islands have ever been discovered which were uninhabited, and animals of a higher grade are as widely diffused as man.

If it would be going too far to claim that all conditions may have forms of life appropriate to them, it would be going as much too far in the other direction to claim that life can exist only with the precise surroundings which nurture it on this planet. It is very remarkable in this connection that while in one direction we see life coming to an end, in the other direction we see it flourishing more and more up to the limit. These two directions are those of heat and cold. We cannot suppose that life would develop in any important degree in a region of perpetual frost, such as the polar regions of our globe. But we do not find any end to it as the climate becomes warmer. On the contrary, everyone knows that the tropics are the most fertile regions of the globe in its production. The luxuriance of the vegetation and the number of the animals continually increase the more tropical the climate becomes. Where the limit may be set no one can say. But it would doubtless be far above the present temperature of the equatorial regions. . . .

We all know that this earth on which we dwell is only one of countless millions of globes scattered through the wilds of infinite space. So far as we know, most of these globes are wholly unlike the earth, being at a temperature so high that, like our sun, they shine by their own light. In such worlds we may regard it as quite certain that no organized life could exist. But evidence is continually increasing that dark and opaque worlds like ours exist and revolve around their suns, as the earth on which we dwell revolves around its central luminary. Although the number of such globes yet discovered is not great, the circumstances under which they are found lead us to believe that the actual number may be as great as that of the visible stars which stud the sky. If so, the probabilities are that millions of them are essentially similar to our own globe. Have we any reason to believe that life exists on these other worlds?

The reader will not expect me to answer this question positively. It must be admitted that, scientifically, we have no light upon the question, and therefore no positive grounds for reaching a conclusion. We can only reason by analogy and by what we know of the origin and conditions of life around us, and assume that the same agencies which are at play here would be found at play under similar conditions in other parts of the universe.

If we ask what the opinion of men has been, we know historically that our race has, in all periods of its history, peopled other regions with beings even higher in the scale of development than we are ourselves. The gods and demons of an earlier age all wielded powers greater than those granted to man—powers which they could use to determine human destiny. But, up to the time that Copernicus showed that the planets were other worlds, the location of these imaginary beings was rather indefinite. It was therefore quite natural that when the moon and planets were found to be dark globes of a size comparable with that of the earth itself, they were made the habitations of beings like unto ourselves.

The trend of modern discovery has been against carrying this view to its extreme, as will be presently shown. Before considering the difficulties in the way of accepting it to the widest extent, let us enter upon some preliminary considerations as to the origin and prevalence of life, so far as we have any sound basis to go upon.

A generation ago the origin of life upon our planet was one of the great mysteries of science. All the facts brought out by investigation into the past history of our earth seemed to show, with hardly the possibility of a doubt, that there was a time when it was a fiery mass, no more capable of serving as the abode of a living being than the interior of a Bessemer steel furnace. There must therefore have been, within a certain period, a beginning of life upon its surface. But, so far as investigation had gone—indeed, so far as it has gone to the present time—no life has been found to originate of itself. The living germ seems to be necessary to the beginning of any living form. Whence, then, came the first germ? Many of our readers may remember a suggestion by Sir William Thomson, now Lord Kelvin, made twenty or thirty years ago, that life may have been brought to our planet by the falling of a meteor from space. This does not, however, solve the difficulty—indeed, it would only make it greater. It still leaves open the question how life began on the meteor; and granting this, why it was not destroyed by the heat generated as the meteor passed through the air. The popular view that life began through a special act of creative power seemed to be almost forced upon man by the failure of science to discover any other beginning for it. It cannot be said that even today anything definite has been actually discovered to refute this view. All we can say about it is that it does not run in with the general views of modern science as to the beginning of things, and that those who refuse to accept it must hold that, under certain conditions which prevail, life begins by a very gradual process, similar to that by which forms suggesting growth seem to originate even under conditions so unfavorable as those existing in a bottle of acid.

But it is not at all necessary for our purpose to decide this question. If life existed through a creative act, it is absurd to suppose that that act was confined to one of the countless millions of worlds scattered through space. If it began at a certain stage of evolution by a natural process, the question will arise, what conditions are favorable to the commencement of this process? Here we are quite justified in reasoning from what, granting this process, has taken place upon our globe during its past history. One of the most elementary principles accepted by the human mind is that like causes produce like effects. The special conditions under which we find life to develop around us may be comprehensively summed up as the existence of water in the liquid form, and the presence of nitrogen, free perhaps in the first place, but accompanied by substances with which it may form combinations. Oxygen, hydrogen, and nitrogen are, then, the fundamental requirements. The addition of calcium or other forms of matter necessary to the existence of a solid world goes without saying. The question now is whether these necessary conditions exist in other parts of the universe.

The spectroscope shows that, so far as the chemical elements go, other worlds are composed of the same elements as ours. Hydrogen especially exists everywhere, and we have reason to believe that the same is true of oxygen and nitrogen. Calcium, the base of lime, is almost universal. So far as chemical elements go, we may therefore take it for granted that the conditions under which life begins are very widely diffused in the universe. It is, therefore, contrary to all the analogies of nature to suppose that life began only on a single world.

It is a scientific inference, based on facts so numerous as not to admit of serious question, that during the history of our globe there has been a continually improving development of life. As ages upon ages pass, new forms are generated, higher in the scale than those which preceded them, until at length reason appears and asserts its sway. In a recent well-known work Alfred Russel Wallace has argued that this development of life required the presence of such a rare combination of conditions that there is no reason to suppose that it prevailed anywhere except on our earth. It is quite impossible in the present discussion to follow his reasoning in detail; but it seems to me altogether inconclusive. Not only does life, but intelligence, flourish on this globe under a great variety of conditions as regards temperature and surroundings, and no sound reason can be shown why under certain conditions, which are frequent in the universe, intelligent beings should not acquire the highest development.

Now let us look at the subject from the view of the mathematical theory of probabilities. A fundamental tenet of this theory is that no matter how improbable a result may be on a single trial, supposing it at all possible, it is sure to occur after a sufficient number of trials—and over and over again if the trials are repeated often enough. For example, if a million grains of corn, of which a single one was red, were all placed in a pile, and a blindfolded person were required to grope in the pile, select a grain, and then put it back again, the chances would be a million to one against his drawing out the red grain. If drawing it meant he should die, a sensible person would give himself no concern at having to draw the grain. The probability of his death would not be so great as the actual probability that he will really die within the next twenty-four hours. And yet if the whole human race were required to run this chance, it is certain that about fifteen hundred, or one out of a million, of the whole human family would draw the red grain and meet his death.

Now apply this principle to the universe. Let us suppose, to fix the ideas, that there are a hundred million worlds, but that the chances are one thousand to one against any one of these taken at random being fitted for the highest development of life or for the evolution of reason. The chances would still be that 100,000 of them would be inhabited by rational beings whom we call human. But where are we to look for these worlds? This no man can tell. We only infer from the statistics of the stars—and this inference is fairly well grounded—that the number of worlds which, so far as we know, may be inhabited, are to be counted by thousands, and perhaps by millions.

In a number of bodies so vast we should expect every variety of conditions as regards temperature and surroundings. If we suppose that the special conditions which prevail on our planet are necessary to the highest forms of life, we still have reason to believe that these same conditions prevail on thousands of other worlds. The fact that we might find the conditions in millions of other worlds unfavorable to life would not disprove the existence of the latter on countless worlds differently situated.

Coming down now from the general question to the specific one, we all know that the only worlds the conditions of which can be made the subject of observation are the planets which revolve around the sun, and their satellites. The question whether these bodies are inhabited is one which, of course, completely transcends not only our powers of observation at present, but every appliance of research that we can conceive of men devising. If Mars is inhabited, and if the people of that planet have equal powers with ourselves, the problem of merely producing an illumination which could be seen in our most powerful telescope would be beyond all the ordinary efforts of an entire nation. An unbroken square mile of flame would be invisible in our telescopes, but a hundred square miles might be seen. We cannot, therefore, expect to see any signs of the works of inhabitants even on Mars. All that we can do is to ascertain with greater or less probability whether the conditions necessary to life exist on the other planets of the system.

The moon being much the nearest to us of all the heavenly bodies, we can pronounce more definitely in its case than in any other. We know that neither air nor water exists on the moon in quantities sufficient to be perceived by the most delicate tests at our command. It is certain that the moon's atmosphere, if any exists, is less than the thousandth part of the density of that around us. The vacuum is greater than any ordinary air-pump is capable of producing. We can hardly suppose that so small a quantity of air could be of any benefit whatever in sustaining life; an animal that could get along on so little could get along on none at all.

But the proof of the absence of life is yet stronger when we consider the results of actual telescopic observation. An object such as an ordinary city block could be detected on the moon. If anything like vegetation were present on its surface, we should see the changes which it would undergo in the course of a month, during one portion of which it would be exposed to the rays of the unclouded sun, and during another to the intense cold of space. If men built cities, or even separate buildings the size of the larger ones on our earth, we might see some signs of them.

In recent times we not only observe the moon with the telescope, but get still more definite information by photography. The whole visible surface has been repeatedly photographed under the best conditions. But no change has been established beyond question, nor does the photograph show the slightest difference of structure or shade which could be attributed to cities or other works of man. To all appearances the whole surface of our satellite is as completely devoid of life as the lava newly thrown from Vesuvius.

We next pass to the planets. Mercury, the nearest to the sun, is in a position very unfavorable for observation from the earth, because when nearest to us it is between us and the sun, so that its dark hemisphere is presented to us. Nothing satisfactory has yet been made out as to its condition. We cannot say with certainty whether it has an atmosphere or not. What seems very probable is that the temperature on its surface is higher than any of our earthly animals could sustain. But this proves nothing.

We know that Venus has an atmosphere. This was very conclusively shown during the transits of Venus in 1874 and 1882. But this atmosphere is so filled with clouds or vapor that it does not seem likely that we ever get a view of the solid body of the planet through it. Some observers have thought they could see spots on Venus day after day, while others have disputed this view. On the whole, if intelligent inhabitants live there, it is not likely that they ever see sun or stars. Instead of the sun they see only an effulgence in the vapory sky which disappears and reappears at regular intervals.

When we come to Mars, we have more definite knowledge, and there seem to be greater possibilities for life there than in the case of any other planet besides the earth. The main reason for denying that life such as ours could exist there is that the atmosphere of Mars is so rare that, in the light of the most recent researches, we cannot be fully assured that it exists at all. The very careful comparisons of the spectra of Mars and of the moon made by Campbell at the Lick Observatory failed to show the slightest difference in the two. If Mars had an atmosphere as dense as ours, the result could be seen in the darkening of the lines of the spectrum produced by the double passage of the light through it. There were no lines in the spectrum of Mars that were not seen with equal distinctness in that of the moon. But this does not prove the entire absence of an atmosphere. It only shows a limit to its density. It may be one-fifth or one-fourth the density of that on the earth, but probably no more.

That there must be something in the nature of vapor at least seems to be shown by the formation and disappearance of the white polar caps of this planet. Every reader of astronomy at the present time knows that, during the Martian winter, white caps form around the pole of the planet which is turned away from the sun, and grow larger and larger until the sun begins to shine upon them, when they gradually grow smaller, and perhaps nearly disappear. It seems, therefore, fairly well proved that, under the influence of cold, some white substance forms around the polar regions of Mars which evaporates under the influence of the sun's rays. It has been supposed that this substance is snow, produced in the same way that snow is produced on the earth, by the evaporation of water.

But there are difficulties in the way of this explanation. The sun sends less than half as much heat to Mars as to the earth, and it does not seem likely that the polar regions can ever receive enough of heat to melt any considerable quantity of snow. Nor does it seem likely that any clouds from which snow could fall ever obscure the surface of Mars.

But a very slight change in the explanation will make it tenable. Quite possibly the white deposits may be due to something like hoar-frost condensed from slightly moist air, without the actual production of snow. This would produce the effect that we see. Even this explanation implies that Mars has air and water, rare though the former may be. It is quite possible that air as thin as that of Mars would sustain life in some form. Life not totally unlike that on the earth may therefore exist upon this planet for anything that we know to the contrary. More than this we cannot say.

In the case of the outer planets the answer to our question must be in the negative. It now seems likely that Jupiter is a body very much like our sun, only that the dark portion is too cool to emit much, if any, light. It is doubtful whether Jupiter has anything in the nature of a solid surface. Its interior is in all likelihood a mass of molten matter far above a red heat, which is surrounded by a comparatively cool, yet, to our measure, extremely hot, vapor. The belt-like clouds which surround the planet are due to this vapor combined with the rapid rotation. If there is any solid surface below the atmosphere that we can see, it is swept by winds such that nothing we have on earth could withstand them. But, as we have said, the probabilities are very much against there being anything like such a surface. At some great depth in the fiery vapor there is a solid nucleus; that is all we can say.

The planet Saturn seems to be very much like that of Jupiter in its composition. It receives so little heat from the sun that, unless it is a mass of fiery vapor like Jupiter, the surface must be far below the freezing-point.

We cannot speak with such certainty of Uranus and Neptune; yet the probability seems to be that they are in much the same condition as Saturn. They are known to have very dense atmospheres, which are made known to us only by their absorbing some of the light of the sun. But nothing is known of the composition of these atmospheres.

To sum up our argument: the fact that, so far as we have yet been able to learn, only a very small proportion of the visible worlds scattered through space are fitted to be the abode of life does not preclude the probability that among hundreds of millions of such worlds a vast number are so fitted. Such being the case, all the analogies of nature lead us to believe that, whatever the process which led to life upon this earth—whether a special act of creative power or a gradual course of development—through that same process does life begin in every part of the universe fitted to sustain it. The course of development involves a gradual improvement in living forms, which by irregular steps rise higher and higher in the scale of being. We have every reason to believe that this is the case wherever life exists. It is, therefore, perfectly reasonable to suppose that beings, not only animated, but endowed with reason, inhabit countless worlds in space. It would, indeed, be very inspiring could we learn by actual observation what forms of society exist throughout space, and see the members of such societies enjoying themselves by their warm firesides. But this, so far as we can now see, is entirely beyond the possible reach of our race, so long as it is confined to a single world.

CHAPTER FIVE

Constellation and Star Names

"Constellation and Star Names" appeared as Chapter 3 in Newcomb's book The Stars: A Study of the Universe *(1901). Here Newcomb tells the story of how astronomers ancient and modern came to divide the sky up into dozens of imaginary figures, and how that division, in turn, came to serve as the basis for the mapping and naming of the stars. Alas, the would-be stargazer of the twenty-first century will, in all likelihood, have to go a good deal further than yard or housetop to obtain a good view of (to paraphrase Newcomb) the starry vault in all its sublimity. For most of us, light pollution has drawn the curtain on what, even a hundred years ago in urban North America, remained a shared experience of all humanity dating back to our earliest ancestors.*

> *Now came still evening on, and twilight grey*
> *Had in her sober livery all things clad.*
> *. . . now glowed the firmament*
> *With living sapphires; Hesperus that led*
> *The starry host rode brightest.* —MILTON

It is strongly recommended to the reader to study the constellations for himself. If he desires to feel all the sublimity associated with them, he must not be satisfied with the hurried glance or occasional survey to which one commonly confines himself in his evening walk. What he should do is, on a clear and moonless summer evening, to escape from his usual surroundings, and go to a place, whether field or housetop, where there is nothing to obstruct his vision, or disturb the current of his thoughts. There he must recline on his back, so as to take in as much as possible of the starry vault at one view. One doing this for the first time will be surprised at the magnificence of the spectacle. As he looks upon the "universal frame" and reflects that it has stood as he now sees it through ages compared with which the whole period of human history is but a fleeting moment, the mind will be filled with a consciousness of infinity and eternity which never before entered it. Other sights become stale from custom, but this can never lose its relish. It can be enjoyed without knowing the name of a constellation, but is more impressive when one reflects that the eyes of man have gazed upon and studied it ever since our race appeared on earth.

In ancient times the practice was adopted of imagining the figures of heroes and animals to be so outlined in the heavens as to include in each figure a large group of the brighter stars. In a few cases some vague resemblance may be traced between the configurations of the stars and the features of the object they are supposed to represent; in general, however, the object chosen seems quite arbitrary. One animal or man could be fitted in as well as another. There is no historic record as to the time when the constellations were mapped out, or of the process by which the outlines were traced. The names of heroes, such as *Perseus*, *Cepheus*, *Hercules*, etc., intermingled with the names of goddesses, show that the time was probably during the heroic age. No maps are extant showing exactly how each figure was placed in the constellation; but in the catalogue of stars given by Ptolemy in his *Almagest*, the positions of particular stars on the supposed body of the hero, goddess, or animal are designated. For example, Aldebaran is said to have formed the eye of the Bull. Two stars marked the right and left shoulders of Orion, and a small cluster marked the position of his head. A row of three stars in a horizontal line showed his belt, three stars in a vertical line below them his sword. From these statements the position of the figure can be reproduced with a fair degree of certainty.

In the well-known constellation *Ursa Major*, the Great Bear, familiarly known as "the Dipper," three stars form the tail of the animal, and four others a part of his body. This formation is not unnatural, yet the figure of a dipper fits the stars much better than that of a bear. In *Cassiopeia*, which is on the opposite side of the pole from the Dipper, the brighter stars may easily be imagined to form a chair in which a lady may be seated. As a general rule, however, the resemblances of the stars to the figure are so vague that the latter might be interchanged to any extent without detracting from their appropriateness.

In any case, it was impossible so to arrange the figures that they should cover the entire heavens; blank spaces were inevitably left in which stars might be found. In order to include every star in some constellation, the figures have been nearly ignored by modern astronomers, and the heavens have been divided up, by somewhat irregular lines, into patches, each of which contains the entire figure as recognised by ancient astronomers. But all are not agreed as to the exact outlines of these extended constellations, and, accordingly, a star is sometimes placed in one constellation by one astronomer and in another constellation by another.

The confusion thus arising is especially great in the southern hemisphere, where it has been intensified by the subdivision of one of the old constellations. The ancient constellation *Argo* covered so large a region of the heavens, and included so many conspicuous stars, that it was divided into four, representing various parts of a ship—the sail, the poop, the prow, and the hull.

Dr. Gould, while director of the Cordoba Observatory, during the years 1870 to 1880, constructed the *Uranometria Argentina*, in which all the stars visible to the naked eye from the south pole to a parallel of declination ten degrees north of the celestial equator were catalogued and mapped. He made a revision of the boundaries of each constellation in such a way as to introduce greater regularity. The rule generally followed was that the boundaries should, so far as possible, run in either an east-and-west or a north-and-south direction on the celestial sphere. They were so drawn that the smallest possible change should be made in the notation of the conspicuous stars; that is, the rule was that, if possible, each bright star should be in the same constellation as before. The question whether this new division shall replace the ancient one is one on which no consensus of view has yet been reached by astronomers. Simplicity is undoubtedly introduced by Gould's arrangement; yet, in the course of time, owing to precession, the lines on the sphere which now run north and south or east and west will no longer do so, but will deviate almost to any extent. The only advantage then remaining will be that the bounding lines will generally be arcs of great circles.

When the heavens began to be carefully studied, two or three centuries ago, new constellations were introduced by Hevelius and other astronomers to fill the vacant spaces left by the ancient ones of Ptolemy. To some of these rather fantastic names were given; the *Bull of Poniatowski*, for example. Some of these new additions have been retained to the present time, but in other cases the space occupied by the proposed new constellation was filled up by extending the boundaries of the older ones.

At the present time the astronomical world, by common consent, recognises eighty-nine constellations in the entire heavens. In this enumeration *Argo* is not counted, but its four subdivisions are taken as separate constellations.

A glance at the heavens will make it evident that the problem of designating a star in such a way as to distinguish it from all its neighbours must be a difficult one. If such be the case with the comparatively small number of stars visible to the naked eye, how must it be with the vast number that can be seen only with the telescope? In the case of the great mass of telescopic stars we have no method of designation except by the position of the star and its magnitude; but with the brighter stars, and, indeed, with all that have been catalogued, other means of identification are available.

It is but natural to give a special name to a conspicuous star. That this was done in very early antiquity we know by the allusion to Arcturus in the Book of Job. At least two such names, Castor and Pollux, have come down to us from classical antiquity, but most of the special names given to the stars in modern times are corruptions of certain Arabic designations. As an example we may mention Aldebaran, a corruption of *Al Dabaran*—The Follower. There is, however, a tendency to replace these special names by a designation of the stars on a system devised by Bayer early in the seventeenth century.

This system of naming stars is quite analogous to our system of designating persons by a family name and a Christian name. The family name of a star is that of the constellation to which it belongs. The Christian name is a letter of the Greek or Roman alphabet or a number. As any number of men in different families may have the same Christian name, so the same letter or number may be assigned to stars in any number of constellations without confusion.

The work of Bayer was published under the title of *Uranometria*, of which the first edition appeared in 1601. This work consists mainly of maps of the stars. In marking the stars with letters on the map, the rule followed seems to have been to give the brighter stars the earlier letters in the alphabet. Were this system followed absolutely, the brightest star should always be called Alpha; the next in order Beta, etc. But this is not always the case. Thus in the constellation *Gemini*, the brightest star is Pollux, which is marked Beta, while Alpha is the second brightest. What system, if any, Bayer adopted in detail has been a subject of discussion, but does not appear to have been satisfactorily made out. Quite likely Bayer himself did not attempt accurate observations on the brightness of the stars, but followed the indications given by Ptolemy or the Arabian astronomers. As the number of stars to be named in several constellations exceeds the number of letters in the Greek alphabet, Bayer had recourse, after the Greek alphabet was exhausted, to letters of the Roman alphabet. In this case the letter *A* was used as a capital, in order, doubtless, that it should not be confounded with the Greek α (alpha). In other cases small italics are used. In several catalogues since Bayer, new italic letters have been added by various astronomers. Sometimes these have met with general acceptance, and sometimes not.

Flamsteed was the first Astronomer Royal of England, and observed at Greenwich from 1666 to 1715. Among his principal works is a catalogue of stars in which the positions are given with greater accuracy than had been attained by his predecessors. He slightly altered the Bayer system by introducing numbers instead of Greek letters. This had the advantage that there was no limit to the number of stars which could be designated in each constellation. He assigned numbers to all the brighter stars in the order of their right ascension, irrespective of the letters used by Bayer. These numbers are extensively used to the present day, and will doubtless continue to be the principal designations of the stars to which they refer. It is very common in our modern catalogues to give both the Bayer letter and the Flamsteed number in the case of Bayer stars.

The catalogues by Flamsteed do not include quite all the stars visible to the naked eye; but various uranometries have been published which were intended to include all such stars. In such cases the designations now used frequently correspond to the numbers given in the uranometries of Bode, Argelander, and Heis.

In recent times these uranometries have been supplemented by censuses of the stars, which are intended to include all the stars to the ninth or tenth magnitude. I shall speak of these in the next section; at present it will suffice to say that stars are very generally designated by their place in such a census.

There is still here and there some confusion both as to the boundaries of the constellations and as to the names of a few of the stars in them. I have already remarked that, in drawing the imaginary boundaries on a star map, as representing the celestial sphere, different astronomers have placed the lines differently. One of the regions in which this is especially true is in the neighbourhood of the north pole, where some astronomers place stars in the constellation *Cepheus* which others place in *Ursa Minor*. Hence in the Bayer system the same star may have different names in different catalogues. Again, in extending the names or numbers, some astronomers use names which others do not regard as authoritative. The remapping of the southern hemisphere by Dr. Gould changed the boundaries of most of the southern constellations in a way already mentioned.

I have spoken of the subdivision of the great constellation *Argo* into four separate ones. Bayer having assigned to the principal stars in this constellation the Greek letters alpha, beta, gamma, etc., the general practice among astronomers since the subdivision has been to continue the designation of the stars thus marked as belonging to the constellation *Argo*. Thus, for example, we have *Alpha Argus*, which after the subdivision belonged to the constellation *Carina*. The variable star *Eta Argus* also belongs to the constellation *Carina*. But in the case of stars not marked by Bayer, the names were assigned according to the subdivided constellations, *Vela*, *Carina*, etc. Confusing though this proceeding may appear to be, it is not productive of serious trouble. The main point is that the same star should always have the same name in successive catalogues. Still, however, it has recently become quite common to ignore the constellation *Argo* altogether and use only the names of its subdivisions. The reader must therefore be on his guard against any mistake arising in this way in the study of astronomical literature.

In star catalogues the position of a star in the heavens is sometimes given in connection with its name. In this case the confusion arising from the same star having different names may be avoided, since a star can always be identified by its right ascension and declination. The fact is that, so far as mere identification is concerned, nothing but the statement of a star's position is really necessary. Unfortunately, the position constantly changes through the precession of the equinoxes, so that this designation of a star is a variable quantity. Hence the special names which we have described are the most convenient to use in the case of well-known stars. In other cases a star is designated by its number in some well-known catalogue. But even here different astronomers choose different catalogues, so that there are still different designations for the same star. The case is one in which uniformity of practice is unattainable.

CHAPTER SIX

The Universe as an Organism

The size and shape of the Milky Way galaxy and our sun's location within the Milky Way were not established until well into the twentieth century. Until then, astronomers believed the solar system was located close to the center of the Milky Way. As it turned out, vast clouds of dust and gas hide the actual center from sight: we are well off to one side, in one of the galaxy's quieter suburbs. In the following essay, Newcomb sets out the prevailing turn-of-the-century view that the sun was near the galaxy's center, but devotes much of the piece to what he calls "stellar statistics," the study of the velocities of the stars. It was, in fact, further work in this area (as well as other hard-won evidence) that eventually pointed the way to a correct understanding of the sun's location. Also noteworthy is Newcomb's anticipation of the importance of "newly discovered emanations . . . or forms of force" in shaping the universe. During the twentieth century, the study of these fundamental forces would prove, as Newcomb predicted, virtually "illimitable" in its scope and implications.

"The Universe as an Organism" was first presented as an address to the Astronomical and Astrophysical Society of America (the body now known as the American Astronomical Society) on 29 December 1902 and was published in Science *the next month. Newcomb was the Society's first president, serving in that capacity from 1899 to 1905.*

I f I were called upon to convey, within the compass of a single sentence, an idea of the trend of recent astronomical and physical science, I should say that it was in the direction of showing the universe to be a connected whole. The farther we advance in knowledge, the clearer it becomes that the bodies which are scattered through the celestial spaces are not completely independent existences, but have, with all their infinite diversity, many attributes in common.

In this we are going in the direction of certain ideas of the ancients which modern discovery long seemed to have contradicted. In the infancy of the race, the idea that the heavens were simply an enlarged and diversified earth, peopled by beings who could roam at pleasure from one extreme to the other, was a quite natural one. The crystalline sphere or spheres which contained all formed a combination of machinery revolving on a single plan. But all bonds of unity between the stars began to be weakened when Copernicus showed that there were no spheres, that the planets were isolated bodies, and that the stars were vastly more distant than the planets. As discovery went on and our conceptions of the universe were enlarged, it was found that the system of the fixed stars was made up of bodies so vastly distant and so completely isolated that it was difficult to conceive of them as standing in any definable relation to one another. It is true that they all emitted light, else we could not see them, and the theory of gravitation, if extended to such distances, a fact not then proved, showed that they acted on one another by their mutual gravitation. But this was all. Leaving out light and gravitation, the universe was still, in the time of Herschel, composed of bodies which, for the most part, could not stand in any known relation one to the other.

When, forty years ago, the spectroscope was applied to analyze the light coming from the stars, a field was opened not less fruitful than that which the telescope made known to Galileo. The first conclusion reached was that the sun was composed almost entirely of the same elements that existed upon the earth. Yet, as the bodies of our solar system were evidently closely related, this was not remarkable. But very soon the same conclusion was, to a limited extent, extended to the fixed stars in general. Such elements as iron, hydrogen, and calcium were found not to belong merely to our earth, but to form important constituents of the whole universe. We can conceive of no reason why, out of the infinite number of combinations which might make up a spectrum, there should not be a separate kind of matter for each combination. So far as we know, the elements might merge into one another by insensible gradations. It is, therefore, a remarkable and suggestive fact when we find that the elements which make up bodies so widely separate that we can hardly imagine them having anything in common, should be so much the same.

In recent times what we may regard as a new branch of astronomical science is being developed, showing a tendency towards unity of structure throughout the whole domain of the stars. This is what we now call the science of stellar statistics. The very conception of such a science might almost appall us by its immensity. The widest statistical field in other branches of research is that occupied by sociology. Every country has its census, in which the individual inhabitants are classified on the largest scale and the combination of these statistics for different countries may be said to include all the interest of the human race within its scope. Yet this field is necessarily confined to the surface of our planet. In the field of stellar statistics millions of stars are classified as if each taken individually were of no more weight in the scale than a single inhabitant of China in the scale of the sociologist. And yet the most insignificant of these suns may, for aught we know, have planets revolving around it, the interests of whose inhabitants cover as wide a range as ours do upon our own globe.

The statistics of the stars may be said to have commenced with Herschel's gauges of the heavens, which were continued from time to time by various observers, never, however, on the largest scale. The subject was first opened out into an illimitable field of research through a paper presented by Kapteyn to the Amsterdam Academy of Sciences in 1893. The capital results of this paper were that different regions of space contain different kinds of stars and, more especially, that the stars of the Milky Way belong, in part at least, to a different class from those existing elsewhere. Stars not belonging to the Milky Way are, in large part, of a distinctly different class.

The outcome of Kapteyn's conclusions is that we are able to describe the universe as a single object, with some characters of an organized whole. A large part of the stars which compose it may be considered as divisible into two groups. One of these comprises the stars composing the great girdle of the Milky Way. These are distinguished from the others by being bluer in color, generally greater in absolute brilliancy, and affected, there is some reason to believe, with rather slower proper motions. The other classes are stars with a greater or less shade of yellow in their color, scattered through a spherical space of unknown dimensions, but concentric with the Milky Way. Thus a sphere with a girdle passing around it forms the nearest approach to a conception of the universe which we can reach today. The number of stars in the girdle is much greater than that in the sphere.

The feature of the universe which should therefore command our attention is the arrangement of a large part of the stars which compose it in a ring, seemingly alike in all its parts, so far as general features are concerned. So far as research has yet gone, we are not able to say decisively that one region of this ring differs essentially from another. It may, there-fore, be regarded as forming a structure built on a uniform plan throughout.

All scientific conclusions drawn from statistical data require a critical investigation of the basis on which they rest. If we are going, from merely counting the stars, observing their magnitudes and determining their proper motions, to draw conclusions as to the structure of the universe in space, the question may arise how we can form any estimate whatever of the possible distance of the stars, a conclusion as to which must be the very first step we take. We can hardly say that the parallaxes of more than one hundred stars have been measured with any approach to certainty. The individuals of this one hundred are situated at very different distances from us. We hope, by long and repeated observations, to make a fairly ap-proximate determination of the parallaxes of all the stars whose distance is less than twenty times that of Alpha Centauri. But how can we know anything about the distance of stars outside this sphere? What can we say against the view of Kepler that the space around our sun is very much thinner in stars than it is at a greater distance; in fact, that the great mass of the stars may be situated between the surfaces of two concentrated spheres not very different in radius. May not this universe of stars be somewhat in the nature of a hollow sphere?

This objection requires very careful consideration on the part of all who draw conclusions as to the distribution of stars in space and as to the ex-tent of the visible universe. The steps to a conclusion on the subject are briefly these: First, we have a general conclusion, the basis of which I have already set forth, that, to use a loose expression, there are likenesses throughout the whole diameter of the universe. There is, therefore, no rea-son to suppose that the region in which our system is situated differs in any essential degree from any other region near the central portion. Again, spectroscopic examinations seem to show that all the stars are in motion, and that we cannot say that those in one part of the universe move more rapidly than those in another. This result is of the greatest value for our purpose, because, when we consider only the apparent motions, as ordi-narily observed, these are necessarily dependent upon the distance of the star. We cannot, therefore, infer the actual speed of a star from ordinary observations until we know its distance. But the results of spectroscopic measurements of radial velocity are independent of the distance of the star.

But let us not claim too much. We cannot yet say with certainty that the stars which form the agglomerations of the Milky Way have, beyond doubt, the same average motion as the stars in other regions of the universe. The difficulty is that these stars appear to us so faint individually, that the investigation of their spectra is still beyond the powers of our instruments. But the extraordinary feat performed at the Lick Observatory of measuring the radial motion of 1830 Groombridge, a star quite invisible to the naked eye, and showing that it is approaching our system with a speed of between fifty and sixty miles a second, may lead us to hope for a speedy solution of this question. But we need not await this result in order to reach very probable conclusions. The general outcome of researches on proper motions tends to strengthen the conclusions that the Keplerian sphere, if I may use this expression, has no very well marked existence. The laws of stellar velocity and the statistics of proper motions, while giving some color to the view that the space in which we are situated is thinner in stars than elsewhere, yet show that, as a general rule, there are no great agglomerations of stars elsewhere than in the region of the Milky Way.

With unity there is always diversity; in fact, the unity of the universe on which I have been insisting consists in part of diversity. It is very curious that, among the many thousands of stars which have been spectroscopically examined, no two are known to have absolutely the same physical constitution. It is true that there are a great many resemblances. Alpha Centauri, our nearest neighbor, if we can use such a word as "near" in speaking of its distance, has a spectrum very like that of our sun, and so has Capella. But even in these cases careful examination shows differences. These differences arise from variety in the combinations and temperature of the substances of which the star is made up. Quite likely also, elements not known on the earth may exist on the stars, but this is a point on which we cannot yet speak with certainty.

Perhaps the attribute in which the stars show the greatest variety is that of absolute luminosity. One hundred years ago it was naturally supposed that the brighter stars were the nearest to us, and this is doubtless true when we take the general average. But it was soon found that we cannot conclude that because a star is bright, therefore it is near. The most striking example of this is afforded by the absence of measurable parallaxes in the two bright stars, Canopus and Rigel, showing that these stars, though of the first magnitude, are immeasurably distant. A remarkable fact is that these conclusions coincide with that which we draw from the minuteness of the proper motions. Rigel has no motion that has certainly been shown by more than a century of observation, and it is not certain that Canopus has either. From this alone we may conclude, with a high degree of probability, that the distance of each is immeasurably great. We may say with certainty that the brightness of each is thousands of times that of the sun, and with a high degree of probability that it is hundreds of thousands of times. On the other hand, there are stars comparatively near us of which the light is not the hundredth part of the sun.

The universe may be a unit in two ways. One is that unity of structure to which our attention has just been directed. This might subsist forever without one body influencing another. The other form of unity leads us to view the universe as an organism. It is such by mutual action going on between its bodies. A few years ago we could hardly suppose or imagine that any other agents than gravitation and light could possibly pass through spaces so immense as those which separate the stars.

The most remarkable and hopeful characteristic of the unity of the universe is the evidence which is being gathered that there are other agencies whose exact nature is yet unknown to us, but which do pass from one heavenly body to another. The best established example of this yet obtained is afforded in the case of the sun and the earth.

The fact that the frequency of magnetic storms goes through a period of about eleven years, and is proportional to the frequency of sunspots, has been well established. The recent work of Professor Bigelow shows the coincidence to be of remarkable exactness, the curves of the two phenomena being practically coincident so far as their general features are concerned. The conclusion is that spots on the sun and magnetic storms are due to the same cause. This cause cannot be any change in the ordinary radiation of the sun, because the best records of temperature show that, to whatever variations the sun's radiation may be subjected, they do not change in the period of the sunspots. To appreciate the relation, we must recall that the researches of Hale with the spectro-heliograph show that spots are not the primary phenomenon of solar activity, but are simply the outcome of processes going on constantly in the sun which result in spots only in special regions and on special occasions. It does not, therefore, necessarily follow that a spot does cause a magnetic storm. What we should conclude is that the solar activity which produces a spot also produces the magnetic storm.

When we inquire into the possible nature of these relations between solar activity and terrestrial magnetism, we find ourselves so completely in the dark that the question of what is really proved by the coincidence may arise. Perhaps the most obvious explanation of fluctuations in the earth's magnetic field to be inquired into would be based on the hypothesis that the space through which the earth is moving is in itself a varying magnetic field of vast extent. This explanation is tested by inquiring whether the fluctuations in question can be explained by supposing a disturbing force which acts substantially in the same direction all over the globe. But a very obvious test shows that this explanation is untenable. Were it the correct one, the intensity of the force in some regions of the earth would be diminished and in regions where the needle pointed in the opposite direction would be increased in exactly the same degree. But there is no relation traceable either in any of the regular fluctuations of the magnetic force, or in those irregular ones which occur during a magnetic storm. If the horizontal force is increased in one part of the earth, it is very apt to show a simultaneous increase the world over, regardless of the direction in which the needle may point in various localities. It is hardly necessary to add that none of the fluctuations in terrestrial magnetism can be explained on the hypothesis that either the moon or the sun acts as a magnet. In such a case the action would be substantially in the same direction at the same moment the world over.

Such being the case, the question may arise whether the action producing a magnetic storm comes from the sun at all, and whether the fluctuations in the sun's activity and in the earth's magnetic field may not be due to some cause external to both. All we can say in reply to this is that every effort to find such a cause has failed and that it is hardly possible to imagine any cause producing such an effect. It is true that the solar spots were, not many years ago, supposed to be due in some way to the action of the planets. But, for reasons which it would be tedious to go into at present, we may fairly regard this hypothesis as being completely disproved. There can, I conclude, be little doubt that the eleven-year cycle of change in the solar spots is due to a cycle going on in the sun itself. Such being the case, the corresponding change in the earth's magnetism must be due to the same cause.

We may, therefore, regard it as a fact sufficiently established to merit further investigation that there does emanate from the sun, in an irregular way, some agency adequate to produce a measurable effect on the magnetic needle. We must regard it as a singular fact that no observations yet made give us the slightest indication as to what this emanation is. The possibility of defining it is suggested by the discovery within the past few years that, under certain conditions, heated matter sends forth entities known as Röntgen rays, Becquerel corpuscles and electrons. I cannot speak authoritatively on this subject, but, so far as I am aware, no direct evidence has yet been gathered showing that any of these entities reach us from the sun. We must regard the search for the unknown agency so fully proved as among the most important tasks of the astronomical physicist of the present time. From what we know of the history of scientific discovery, it seems highly probable that, in the course of his search, he will, before he finds the object he is aiming at, discover many other things of equal or greater importance of which he had, at the outset, no conception.

The main point I desire to bring out in this review is the tendency which it shows towards unification in physical research. Heretofore differentiation—the subdivision of workers into a continually increasing number of groups of specialists—has been the rule. Now we see a coming together of what, at first sight, seem the most widely separated spheres of activity. What two branches could be more widely separated than that of stellar statistics, embracing the whole universe within its scope, and the study of these newly discovered emanations, the product of our laboratories, which seem to show the existence of corpuscles smaller than the atoms of matter? And yet, the phenomena which we have reviewed, especially the relation of terrestrial magnetism to the solar activity, and the formation of nebulous masses around the new stars, can be accounted for only by emanations or forms of force, having probably some similarity with the corpuscles, electrons, and rays which we are now producing in our laboratories. The nineteenth century, in passing away, points with pride to what it has done. It has become a word to symbolize what is most important in human progress. Yet, perhaps its greatest glory may prove to be that the last thing it did was to lay a foundation for the physical science of the twentieth century. What shall be discovered in the new fields is, at present, as far without our ken as were the modern developments of electricity without the ken of the investigators of one hundred years ago. We cannot guarantee any special discovery. What lies before us is an illimitable field, the existence of which was scarcely suspected ten years ago, the exploration of which may well absorb the activities of our physical laboratories, and of the great mass of our astronomical observers and investigators, for as many generations as were required to bring electrical science to its present state. We of the older generation cannot hope to see more than the beginning of this development, and can only tender our best wishes and most hearty congratulations to the younger school whose function it will be to explore the limitless field now before it.

CHAPTER SEVEN

The Coming Total Eclipse of the Sun

Newcomb observed solar eclipses in Iowa in 1869, Gibraltar in 1870, and in Wyoming in 1878. He was, therefore, well equipped to provide a preview of a total eclipse which was to be visible from parts of the southern United States on May 28, 1900, and did so, both in the following article, which appeared in McClure's Magazine *in May 1900, as well as in a piece for the New York* Evening Post *on May 19. Beginning with a compelling description of what it's like to observe a total eclipse, Newcomb goes on to consider the phenomenon's scientific importance. And, in fact, the May 1900 eclipse was scientifically important, for it resulted in ground-breaking photography of the solar corona by Thomas William Smillie of the Smithsonian Institution; one of Smillie's photographs is reproduced below.*

A total eclipse of the sun is one of the most impressive sights that nature offers to the eye of man. To see it to the best advantage, one should be in an elevated position commanding the largest possible view of the surrounding country, especially in the direction from which the shadow of the moon is to come. The first indication of anything unusual is to be seen, not on the earth or in the air, but on the disk of the sun. At the predicted moment, a little notch will be seen to form somewhere on the western edge of the sun's outline. It increases minute by minute, gradually eating away as it were the visible sun. No wonder that imperfectly civilized people, when they saw the great luminary thus diminishing in size, fancied that a dragon was devouring its substance.

For some time, perhaps an hour, nothing will be noticed but the continued progress of the advancing moon. It will be interesting if, during this time, the observer is in the neighborhood of a tree that will permit the sun's rays to reach the ground through the small openings in its foliage. The little images of the sun which form here and there on the ground will then have the form of the partially eclipsed sun. Soon the latter appears as the new moon, only instead of increasing, the crescent form grows thinner minute by minute. Even then, so well has the eye accommodated itself to the diminishing light, there may be little noticeable darkness until the crescent has grown very thin. If the observer has a telescope with a dark glass for viewing the sun, he will now have an excellent opportunity of seeing the mountains on the moon. The unbroken limb of the sun will keep its usual soft and uniform outline. But the inside of the crescent, the edge of which is formed by the surface of the moon, will be rough and jagged in outline.

A few minutes before the last vestige of the sun is to disappear, the growing darkness will become very noticeable. It is a curious fact that the darkness does not seem to come on uniformly, but like a series of shadows, following each other at intervals of a few seconds. The cause of these seeming shadows has been the subject of some discussion; but there is reason to believe that they are an optical illusion, caused by the unequal rate at which the eye accommodates itself to the diminution of light.

Eclipse of May 28, 1900, showing the corona. Photograph by Thomas William Smillie of the Smithsonian Institution. (Smithsonian Institution Archives)

A short time before the fading crescent is to disappear, the observer should look toward the point from which the shadow is to come—commonly not far from the west, say between southwest and northwest. If the air is quite clear, the shadow will first be seen on the distant horizon, advancing at the rate of a mile in every two, three, or four seconds, according to circumstances. The nearer the time is to noon, the slower will be the advance, and the more impressive the sight. On it comes. In a moment the spectator will be enveloped in it. The advancing mountains on the rugged surface of the moon have reached the sun's edge, and nothing is seen of the latter except a row of broken fragments or points of light, shining between the hollows on the lunar surface. They last but a second or two before they vanish.

Now is seen the glory of the spectacle. The sky is clear and the sun in mid-heaven, and yet no sun is visible. Where the latter ought to be, the densely black globe of the moon hangs, as it were, in mid-air. It is surrounded by an effulgence radiating a saintly glory. This is now known as the "corona." Though bright enough to the unaided vision, it is seen to the best advantage with a telescope of very low magnifying power. Even a common opera glass may suffice. With a telescope of high power only a portion of the corona is visible, and thus the finest part of the effect is lost. A common spy-glass, magnifying ten or twelve times, is better, so far as the splendor of the effect is concerned, than the largest telescope. Such an instrument will show, not only the corona itself, but the so-called "prominences"—fantastic cloud-like forms of rosy color rising here and there, seemingly from the dark body of the moon.

The darkness during the height of an eclipse is not so great as it is sometimes supposed to be. The sun still illuminates the atmosphere outside the region of the shadow, casting into the whole dark interior a "disastrous twilight," as Milton calls it, strong enough to enable the astronomer to read the time by his chronometer without difficulty. It may be likened to the actual twilight about half an hour after sunset.

Under any circumstances the observer will have but a short time to enjoy the scene. In a minute or two, perhaps three, four, or five minutes, according to circumstances, sunlight will be seen coming from the same direction as that from which the shadow advanced. A few seconds more, and it flashes upon the observer. The glory disappears in a moment, and, except for the partially eclipsed sun, nature assumes her usual aspect.

Much has been written about the effect of such an eclipse upon animals. Quite likely these descriptions have been exaggerated. But it has not always been thus in the case of men. Arago tells of a girl in the south of France who was tending cattle in the fields during the eclipse of 1842, which was total over the region in which she lived. Filled with alarm at the black object which had usurped the place of the sun, she ran forward crying. When light returned and the sun reappeared, she dried her tears with the exclamation, "Oh, beautiful sun!"

Of late years a powerful aid has been lent to astronomy by photography. With the sensitive chemicals now used in the photographic art it is possible to photograph celestial objects which are invisible to the eye. Millions of stars are now being charted in the sky, and thousands of faint nebulae discovered, which the human eye would never have seen, even when aided by the most powerful telescope. Now it is hoped that our astronomers will apply some method of photographing the sky around the sun during the coming eclipse. If there is any object or any group of objects there of which the attraction would produce any effect, we hope that it may be discovered. . . .

Eclipses of the sun are of such general interest to the public, as well as to the astronomer, that the reader will perhaps not be wearied if I say something more about them. A sort of dramatic interest is given to them by the fact, so familiar to all of us, that the sun and moon are almost exactly of the same apparent size. Each of these bodies is at certain times a little nearer to us than at others. When the moon is nearest to us, it seems a little larger than the sun, and when farthest away, a little smaller. It makes all the difference in the world in the character of an eclipse which of these two is the case. In the first case, the moon will entirely hide the sun; in the second it cannot.

To see to the best advantage what will happen, the observer on the earth must choose such a place that the center of the moon will pass exactly over the center of the sun. What he then sees is called a central eclipse. If the moon is a little larger in apparent size, it hides the sun, and the eclipse is total. But if it is smaller, the extreme edge of the sun will be seen all around the dark edge of the moon, forming a ring of sunlight. The eclipse is then annular. Such an eclipse does not offer the same advantage in the study of the sun that a total one does, and is therefore of less scientific interest. But it must be very instructive to anyone who has the opportunity to see it. On the average the apparent size of the moon is smaller than that of the sun, so that annular eclipses occur a little oftener than total ones. . . .

It is remarkable that, though the ancients were familiar with the fact of eclipses, and the more enlightened of them perfectly understood their causes, some even the laws of their recurrence, there are very few actual accounts of these phenomena in the writings of the ancient historians. The old Chinese annals now and then record the fact that an eclipse of the sun occurred at a certain time in some province or near some city of the Empire. But no particulars are given. Quite recently the Assyriologists have deciphered from ancient tablets a statement that an eclipse of the sun was seen at Nineveh, on June 15th, 763 B.C. Our astronomical tables show that there actually was a total eclipse of the sun on this day, during which the shadow passed 100 miles or so north of Nineveh.

Perhaps the most celebrated of the ancient eclipses, and the one that has given rise to most discussion, is that known as the eclipse of Thales. Its principal historical basis is a statement of Herodotus, that in a battle between the Lydians and the Medes the day was suddenly turned into night. The armies thereupon ceased battle and were more eager to come to terms of peace with each other. It is added that Thales, the Milesian, had predicted to the Ionians this change of day, even the very year in which it should occur.

An eclipse of which we have a very explicit statement in the writings of the ancients is now generally known as the eclipse of Agothocles. Agothocles was tyrant of Syracuse, and was long engaged in war with the Carthaginians. In 310 B.C., the latter were blockading his fleet, of which he was in personal command, in the harbor of his own town. He availed himself of a momentary relaxation in the blockade to sail away for the Carthaginian territory. The second day of his voyage, which lasted six days and nights in all, he saw a total eclipse of the sun. This observation would have been of great use to the astronomers of our time in correcting their tables, were they sure of the locality of Agothocles at the time he made it. But it has been an open question whether he sailed directly toward the south or went toward the north, making the circuit of the whole Island of Sicily. The result would be quite different in the two cases. The probability now seems to be that he passed to the north, and this accords with the conclusions from our most recent investigations on the motion of the moon.

In modern times, since it became possible to predict the path of an eclipse along the earth's surface, and the time at which it would begin and end at any given place, the principal interest which astronomers at first took in the phenomenon grew out of the test which it afforded of the tables of the moon's motion. In 1715, the shadow of the moon passed over the western and southeastern parts of England, including London in its range. Halley, who had just been made astronomer royal, planned a more extended and careful series of observations on this eclipse than had ever before been made. Men in various towns near the edge of the shadow noted carefully whether the sun was totally eclipsed or not, and, where it was, how long the total phase lasted. In this way it became possible to lay down on a map, from observations, the limits of the moon's shadow without an error of more than two or three miles. The times of beginning and end of the total phase were also carefully noted in London and its immediate neighborhood.

The French astronomers had a different method of observation, which could be equally well applied whether an eclipse was total or not. They did what any of us can do with the aid of a spyglass: they pointed a telescope at the sun, and then, instead of looking into the telescope, held a screen at some little distance behind it, on which an image of the sun was thrown. By looking at this image the progress of the eclipse could be noted more easily than by looking at the sun itself, because no dark glass was necessary and the observer could sit down and watch the affair at his leisure. . . .

It was not until after the beginning of the nineteenth century that men began to avail themselves of total eclipses to make observations of the sun's surroundings, with a view of throwing light upon the question of the physical constitution of our great luminary. The corona and the prominences had been observed since the seventeenth century, and drawings and descriptions of the appearances made; but it does not seem to have occurred to anyone that questions respecting the nature or cause of these objects could be answered. Even now the reader may inquire how it is that we can learn anything about the sun by hiding him from our sight, and, if we can, why a chimney would not answer the purpose as well as the moon. The answer is not far to seek. In the daytime the whole air around the sun is so brightly illuminated that it is impossible to see anything in the immediate neighborhood of that body. We may cut off the sunlight from our eyes by a chimney, but we cannot cut off the illumination of the air except by an object far above the air. The size and distance of the moon are such that it cuts off a great deal of light for hundreds of miles around us, and enables us to see the region close around the sun through an almost dark sky.

Even when curiosity as to the corona and prominences began to be aroused, it was long before any answers to questions about them were apparent. Anyone could look into a telescope, describe what he saw, and, if a good draftsman, make a picture of the scene. But what could he learn from such a picture? So much in the dark were even the most advanced astronomers on the subject up to the middle of the nineteenth century, that it was not established whether the corona belonged to the sun or to the moon. If, as might be the case, the latter was surrounded by a very rare atmosphere, even one so rare that we could not see it on ordinary occasions, its bright illumination by the rays of the sun might show as a corona around the moon. In 1851 a total eclipse was visible in Northern Europe, which enabled the question of the whereabouts of the red prominences to be settled. It was found that, as the moon traveled along over the sun, she traveled over the prominences also, advancing on those in front, uncovering those behind. This showed that these objects certainly belonged to the sun and not to the moon. The same would probably be true of the corona, but in this case it was difficult to reach so positive a conclusion.

About 1863–64 the spectroscope began to be applied to researches on the heavenly bodies. Mr. (now Sir William) Huggins, of London, was a pioneer in observing the spectra of the stars and nebulas. For several years it did not seem that much was to be learned in this way about the sun. The year 1868 at length arrived. On August 18th there was to be a remarkable total eclipse of the sun, visible in India. The shadow was 140 miles broad; the duration of the total phase was more than six minutes. The French sent Mr. Janssen, one of their leading spectroscopists, to observe the eclipse in India and see what he could find out. Wonderful was his report. The red prominences which had perplexed scientists for two centuries were found to be immense masses of glowing hydrogen, rising here and there from various parts of the sun, of a size compared with which our earth was a mere speck. This was not all. After the sunlight reappeared, Janssen began to watch these objects in his spectroscope. He followed them as more and more of the sun came out, and continued to see them after the eclipse was over. They could be observed at any time when the air was sufficiently clear and the sun high in the sky.

By a singular coincidence this same discovery was made independently in London without any eclipse. Mr. J. Norman Lockyer was then rising into prominence as an enthusiastic worker with the spectroscope. It occurred independently to him and to Mr. Huggins that the heat in the neighborhood of the sun was so intense that any matter that existed there would probably take the form of a gas shining by its own light. The spectrum of such a gas is composed of bright lines, which are but little enfeebled to whatever extent the spectrum as a whole may be spread out by the prism through which it passes. But the sun's light reflected from the air is more and more enfeebled the more it is spread out. Consequently, if a spectroscope of sufficient power were directed at the sun just outside its border, the brilliancy of the light reflected from the air might be so diminished that the bright lines from the gases surrounding the sun would be seen. It was anticipated that thus the prominences would be made visible. Both of the investigators we have mentioned endeavored to get a sight of the prominences in this way; but it was not until October 20th, two months after the Indian eclipse, that Mr. Lockyer succeeded in having an instrument of sufficient power completed. Then, at the first opportunity, he found that he could see the prominences without an eclipse!

At that time communication with India was by mail, so that for the news of Mr. Janssen's discovery astronomers had to wait until a ship arrived. By a singular coincidence his report and Mr. Lockyer's communication announcing his own discovery reached the French Academy of Sciences at the same meeting. This eminent body, with pardonable enthusiasm, caused a medal to be struck in commemoration of the new method of research, in which the profiles of Lockyer and Janssen appeared together as co-discoverers. Since that time the prominences are regularly mapped out from day to day by spectroscopic observers in various parts of the world.

Up to the present time the question of the corona is an unsettled one. There appears to be some yet unsolved mystery enveloping its origin. Everything about it shows that it cannot be an atmosphere of the sun, as was once supposed. Were such the case, it would, unless composed of some substance vastly lighter than hydrogen, be drawn down to the sun's surface by the powerful attraction of that body. It could not rise hundreds of thousands of miles from the sun, as the corona does; and even if it did, its light would be smooth and uniform, whereas the coronal light has a sort of hairy or fibrous structure. This may be seen on most of the good photographs of the corona.

Professor P. H. Bigelow has noticed a remarkable resemblance between these seeming fibers and the curves which iron filings scattered over paper assume when we place a magnet under the paper. He has thus formed a theory of the corona based on some action of the sun akin to magnetism. The coincidence between the results of this theory and the general figure of the corona, especially the direction of the fibers, is, to say the least, very curious. Some sort of polarization in the direction of the sun's axis seems to be clearly indicated. But we have here no explanation as to how the matter forming the corona is kept from falling into the sun by the powerful attraction of gravity, which is there twenty-seven times what it is on the earth. Quite likely this is brought about by some form of electrical or other repulsion, similar to that which is seen to act in the tail of a comet.

Another mystery is the nature of the long streamers, sometimes extending far beyond the outer parts of the corona. Some analogy has been suspected between these and the streamers of the aurora. The view has thus arisen that the corona may be an aurora around the sun. More observations and studies must be made, both upon the aurora and the phenomena of terrestrial magnetism, before we can reach any decided conclusion on this question.

The composite nature of the spectrum of the corona shows that the substance which forms it is not all in the same state. Most of the light which it emits gives an unbroken spectrum, seemingly without dark lines. This shows that it emanates partly from hot particles, and not wholly from diffusing gases. It is likely that this matter shines partly by its own light and partly by the reflected light of the sun. But there are also bright lines in the spectrum, one of which has particularly attracted the attention of investigators ever since its discovery in 1869. It seems to be emitted by some gas not known to exist upon the surface of the earth, and to which the name *coronium* has been given. It is interesting to remark in this connection that the solar spectrum shows at least one other substance in the sun which was formerly not known to exist on the earth, and which was therefore called *helium*. But, only a few years ago, this substance was found in clevite, a somewhat rare mineral of Norway. Possibly we may yet discover coronium somewhere on the earth.

We may consider it as certain that the corona, considered as a mass of matter, is a very flimsy affair. When we recall that its extent is to be measured by hundreds of thousands, nay, millions of miles, and that it surrounds a globe of more than a hundred times the diameter of the earth, and therefore having more than ten thousand times the earth's surface, we might think of it as a very massive structure. But we should be deceived. A few quarts of water condensed in the air will make a very respectable-looking fog or cloud. Such a cloud in the immediate neighborhood of the sun would shine with a hundred thousand times the light which any terrestrial cloud ever shone with in the brightest rays of the sun. Quite likely, if we should surround the earth with a corona like that of the sun, we should never be able to see it, or to detect its existence in the air or above the air, by any research we could make. But an observer on the moon would see it plainly. It would be the same with the tail of a comet, which is so tenuous that we can see a small star through a million miles of its thickness. Fifty miles' thickness would not suffice to make it visible in the brightest rays of the sun.

Perhaps the most interesting object which the spectroscopists have examined during total eclipses is known as the "reversing layer." This was first discovered by Professor Young, during the eclipse of 1870, which he observed in Spain. He was noticing the changes in the appearance of the spectrum given by the sun's light when the moon was nearly cutting it off. At the very last moment, when no part of the sun was visible except its extreme edge, the dark lines of the spectrum were changed to bright ones.

As the last ray disappeared, all the bright lines of the spectrum flashed out. This showed that the substances which compose the sun exist at its immediate surface as a layer of glowing gases, all substances being vaporized by the fervent heat which there prevails. This heat is more intense than anything we can produce by terrestrial means.

The questions that relate to the sun are not the only ones that total eclipses enable the astronomer to attack. Such of our readers as have specially interested themselves in celestial science are doubtless aware that the motion of the planet Mercury shows a minute deviation which might be produced by the attraction of a planet, or group of planets, between it and the sun. This deviation was first discovered by Le Verrier, celebrated as having computed the position of Neptune before it had ever been recognized in the telescope. His announcement set people to looking for the supposed planet. About 1860, a Dr. Lescarbault, a country physician of France who possessed a small telescope, thought he had seen this planet passing over the disk of the sun. But it was soon proved that he must have been mistaken. Another more experienced astronomer, who was looking at the sun on the same day, failed to see anything except an ordinary spot, which probably misled the physician-astronomer. Now, for forty years the sun has been carefully scrutinized and photographed from day to day at several stations without anything of the sort being seen.

Still, it is possible that little planets so minute as to escape detection in passing over the sun's disk may revolve in the region in question. If so, their light would be completely obscured by that of the sky, so that they might not ordinarily be visible. But there is still a chance that, during a total eclipse of the sun, when the light is cut off from the sky, they could be seen. Observers have, from time to time, looked for them during total eclipses. In one instance something of the sort was supposed to be found. During the eclipse of 1878, Professor Watson, of Ann Arbor, and Professor Lewis Swift, both able and experienced observers, thought that they had detected some such bodies. But critical examination left no doubt that what Watson saw was a pair of fixed stars which had always been in that place. How it was with the observations of Professor Swift has never been certainly ascertained, because he was not able to lay down the position with such certainty that positive conclusions could be drawn. . . .

There is a curious law of recurrence of eclipses, which has been known from ancient times.   It is based on the fact that the sun and moon return to nearly the same positions relative to the node and perigee of the moon's orbit after a period of 6,585 days, 8 hours, or 18 years and 12 days. Hence, eclipses of every sort repeat themselves at this interval. For example, the [eclipse of May 28, 1900] may be regarded as a repetition of those which occurred in the years 1846, 1864, and 1882. But when such an eclipse recurs, it is not visible in the same part of the earth, because of the excess of eight hours in the period. During this eight hours the earth performs one-third of a rotation on its axis, which brings a different part under the sun. Each eclipse is visible in a region about one-third of the way round the world, or 120° of longitude, west of where it occurred before. Only after three periods will the recurrence be near the same region. But in the meantime the moon's line of motion will have changed so that the path of its shadow will pass farther north or south.

A study of the eclipses of the series to which the present one belongs will illustrate the law in question. The first one that we need mention is that of April 25, 1846. The middle point of the shadow-path, that is, the point where the total phase occurred at noon, was then in the West Indies, among the Bahama Islands. This was the first eclipse of the series that was really total, and here it was total only near the middle of the path. The path passed from the Pacific Ocean over northern Mexico, touched the northern end of Cuba, and crossed the Atlantic Ocean to the African coast of the Mediterranean. The central point was in 25° north latitude.

The next recurrence was on May 6, 1864. The shadow swept over the Pacific Ocean, and the middle point of its path was in 32° north latitude. After the lapse of another period, the eclipse returned in 1882, May 17. The shadow swept across the great desert of Sahara, passed through Egypt and the continent of Asia, leaving the earth in the Pacific Ocean south of Japan. The middle point was now in 39° north latitude.

Next we have our present eclipse of May 28th. After passing from New Orleans over the Gulf States along the line shown on our map, the shadow will enter the Atlantic Ocean at Norfolk, cross over to the Spanish Peninsula, and pass along the Mediterranean into Northern Africa. The central point will be in the Atlantic Ocean, in 55° north latitude.

During the three periods of recurrence the changes in the respective positions of the sun and moon have been such as to throw the shade some seven degrees farther north at each recurrence, or about twenty degrees in all. That is, it will now pass twenty degrees farther north than it passed in 1846.

The next period of 6,585 days will bring us to June 8, 1918. The shadow will then pass from near Japan over the northern part of the Pacific Ocean, strike our Pacific coast near the mouth of the Columbia River, and travel over the United States in a southeasterly direction, through Oregon, Idaho, southwest Wyoming, Colorado, Arkansas, and the Gulf States and Florida. Somewhere in Mississippi or Alabama it will cross the path of the present eclipse. At the point of crossing the inhabitants will have the pleasure of seeing two successive total eclipses of the series. Their fortune, however, will not be so remarkable as that of the inhabitants at a point in the Northwest who saw both of the total eclipses of 1869 and 1878.

The series will continue at the regular intervals we have mentioned until August 23, 2044, when the shadow will barely touch the earth in the region of the North Pole. After that it will skip our planet entirely.

There are two series of eclipses remarkable for the long duration of the total phase. To one of these the eclipse of 1868, already mentioned, belongs. This recurred in 1886, and will recur again in 1904. Unfortunately, at the first recurrence, the shadow was cast almost entirely on the Atlantic and Pacific oceans, so that it was not favorable for observation by astronomers. That of September 9, 1904, will be yet more unfortunate for us, because the shadow will pass only over the Pacific Ocean. Possibly, however, it may touch some island where observations may be made. The recurrence of 1922, on September 21, will be visible in Northern Australia, where the duration of totality will be about four minutes.

To the other series belongs the eclipse of 1883. This will recur in 1901, on May 18th, when the moon's shadow will sweep from near Madagascar and cross the Indian Ocean, Sumatra, Borneo, and Papua. Unfortunately, this region is very cloudy, and however carefully the preparations for observations may be made, the astronomers will run a great risk of not seeing the eclipse. But hope springs eternal in the human breast; and it is not likely that observers will be deterred from an heroic attempt by any threats of the weather.

At the successive recurrences the duration of totality will be longer and longer through the twentieth century. In 1937, 1955, and 1973 it will exceed seven minutes, so that, as far as duration is concerned our successors will have a more remarkable opportunity than their ancestors have enjoyed for many centuries.

The question may arise as to the degree of precision with which the path of an eclipse can be predicted by the astronomer. It is sometimes supposed that he can determine a hundred years in advance, and to the exact second, when such a phenomenon will begin or end. This is a great exaggeration of his powers. One entertaining such an idea may have a very high opinion of the power of modern mathematics, but he has no conception of the difficulties of the problem of the moon's motion. The pull which the sun exerts on the earth and moon by its gravitation second after second, minute after minute, hour after hour, day after day, and year after year must be known, and its effect continually added up by a mathematic method of which man had no conception until the time of Leibnitz and Newton. The changes in the positions of the two bodies caused by the pull of the sun continually changes the action of that pull, because, as one will readily see, the latter depends upon the relative positions, while the positions are continually changed by the pull. This is what makes the problem so complicated.

If we had only the sun to deal with, we might hope to get along. But the planets, especially Venus, come in, and insist on having their little pull also. Before their action was found out, there were some deviations in the motion of the moon which are now attributed to the action of Venus. To compute this action is the most complex problem which the mathematical astronomer has to deal with, and he has not yet succeeded in solving it to his satisfaction. And when he has solved it, he is by no means at the end of his trouble. There are several indications that the rotation of the earth slowly changes from time to time, our planet turning on its axis sometimes a little faster and sometimes a little slower. The change is, indeed, very slow; not more than two or three thousandths of a second in a day. But, if it takes to rotating faster even by this minute amount, it will get ahead of its calculated place by a second in a year and a minute in sixty years; and then the astronomer who fixes his point of observation so that he will be carried to exactly the center of the moon's shadow, according to calculations made sixty years before, may find himself out of the way by several miles.

What makes the matter difficult is that these changes in the earth's rotation cannot, so far as we have yet learnt, be exactly observed, or even predicted; they probably arise from changes in the position of ice around the North Pole, changes in ocean currents, and perhaps in the movement of the winds. The reason that they cannot be directly determined is that we cannot make any clock which will keep time year after year without the error of a second. The rotation of the earth on its axis affords the astronomer the only measure of time he can use in his work, and if it goes wrong, he is, for the time being, left at sea. But his motto today is always forward; he has not lost one particle of enthusiasm because his science has been progressing for 2,000 years. He will leave no device untried to learn everything that is to be learned about the motions of the earth and heavenly bodies, confident that if he must fail, his successors will carry on his work to perfection. If today he cannot tell his successors of the year 2000 when to expect an eclipse within one minute by the clock, he of the year 2000 may do it for his successor of 2100.

CHAPTER EIGHT

How the Planets Are Weighed

Generations of textbook authors and popular science writers have struggled to make clear the difference between "mass" and "weight," and in this article, first published in January 1900 in McClure's, *Newcomb does the job as well as anyone. Despite all the advances in astronomy and space science that the twentieth century brought with it, the underlying principle of how we determine the mass of the earth and the other planets remains unchanged, though we know these figures with vastly more accuracy than in Newcomb's day. I know of no subsequent discussion of the topic, however, that employs the analogy of a ham in a butcher shop!*

You ask me how the planets are weighed? I reply, on the same principle by which a butcher weighs a ham in a spring-balance. When he picks the ham up, he feels a pull of the ham towards the earth. When he hangs it on the hook, this pull is transferred from his hand to the spring of the balance. The stronger the pull, the farther the spring is pulled down. What he reads on the scale is the strength of the pull. You know that this pull is simply the attraction of the earth on the ham. But, by a universal law of force, the ham attracts the earth exactly as much as the earth does the ham. So what the butcher really does is to find how much or how strongly the ham attracts the earth, and he calls that pull the weight of the ham. On the same principle, the astronomer finds the weight of a body by finding how strong is its attractive pull on some other body. If the butcher, with his spring-balance and a ham, could fly to all the planets, one after the other, weigh the ham on each, and come back to report the results to an astronomer, the latter could immediately compute the weight of each planet of known diameter, as compared with that of the earth.

In applying this principle to the heavenly bodies, we at once meet a difficulty that looks insurmountable. You cannot get up to the heavenly bodies to do your weighing; how then will you measure their pull? I must begin the answer to this question by explaining a nice point in exact science. Astronomers distinguish between the *weight* of a body and its *mass*. The weight of objects is not the same all over the world; a thing which weighs thirty pounds in New York would weigh an ounce more than thirty pounds in a spring-balance in Greenland, and nearly an ounce less at the equator. This is because the earth is not a perfect sphere, but a little flattened. Thus weight varies with the place. If a ham weighing thirty pounds were taken up to the moon and weighed there, the pull would only be five pounds, because the moon is so much smaller and lighter than the earth. There would be another weight of the ham for the planet Mars, and yet another on the sun, where it would weigh some eight hundred pounds. Hence the astronomer does not speak of the weight of a planet, because that would depend on the place where it was weighed; but he speaks of the mass of the planet, which means how much planet there is, no matter where you might weigh it.

At the same time, we might, without any inexactness, agree that the mass of a heavenly body should be fixed by the weight it would have in New York. As we could not even imagine a planet at New York, because it may be larger than the earth itself, what we are to imagine is this: Suppose the planet could be divided into a million million million equal parts, and one of these parts brought to New York and weighed. We could easily find its weight in pounds or tons. Then multiply this weight by a million million million, and we shall have a weight of the planet. This would be what the astronomers might take as the mass of the planet.

With these explanations, let us see how the weight of the earth is found. The principle we apply is that round bodies of the same specific gravity attract small objects on their surface with a force proportional to the diameter of the attracting body. For example, a body two feet in diameter attracts twice as strongly as one of a foot, one of three feet three times as strongly, and so on. Now, our earth is about 40,000,000 feet in diameter; that is 10,000,000 times four feet. It follows that if we made a little model of the earth four feet in diameter, having the average specific gravity of the earth, it would attract a particle with one ten-millionth part of the attraction of the earth. The attraction of such a model has actually been measured. Since we do not know the average specific gravity of the earth—that being in fact what we want to find out—we take a globe of lead, four feet in diameter, let us suppose. By means of a balance of the most exquisite construction it is found that such a globe does exert a minute attraction on small bodies around it, and that this attraction is a little more than the ten-millionth part of that of the earth. This shows that the specific gravity of the lead is a little greater than that of the average of the whole earth. All the minute calculations made, it is found that the earth, in order to attract with the force it does, must be about five and one-half times as heavy as its bulk of water, or perhaps a little more. Different experimenters find different results; the best between 5.5 and 5.6, so that 5.5 is, perhaps, as near the number as we can now get. This is much more than the average specific gravity of the materials which compose that part of the earth which we can reach by digging mines. The difference arises from the fact that, at the depth of many miles, the matter composing the earth is compressed into a smaller space by the enormous weight of the portions lying above it. Thus, at the depth of 1000 miles, the pressure on every cubic inch is more than 2000 tons, a weight which would greatly condense the hardest metal.

We come now to the planets. I have said that the mass or weight of a heavenly body is determined by its attraction on some other body. There are two ways in which the attraction of a planet may be measured. One is by its attraction on the planets next to it. If these bodies did not attract one another at all, but only moved under the influence of the sun, they would move in orbits having the form of ellipses. They are found to move very nearly in such orbits, only the actual path deviates from an ellipse, now in one direction and then in another, and it slowly changes its position from year to year. These deviations are due to the pull of the other planets, and by measuring the deviations we can determine the amount of the pull, and hence the mass of the planet.

The reader will readily understand that the mathematical processes necessary to get a result in this way must be very delicate and complicated. A much simpler method can be used in the case of those planets which have satellites revolving round them, because the attraction of the planet can be determined by the motions of the satellite. The first law of motion teaches us that a body in motion, if acted on by no force, will move in a straight line. Hence, if we see a body moving in a curve, we know that it is acted on by a force in the direction towards which the motion curves. A familiar example is that of a stone thrown from the hand. If the stone were not attracted by the earth, it would go on forever in the line of throw, and leave the earth entirely. But under the attraction of the earth, it is drawn down and down, as it travels onward, until finally it reaches the ground. The faster the stone is thrown, of course, the farther it will go, and the greater will be the sweep of the curve of its path. If it were a cannon-ball, the first part of the curve would be nearly a right line. If we could fire a cannon-ball horizontally from the top of a high mountain with a velocity of five miles a second, and if it were not resisted by the air, the curvature of the path would be equal to that of the surface of our earth, and so the ball would never reach the earth, but would revolve round it like a little satellite in an orbit of its own. Could this be done, the astronomer would be able, knowing the velocity of the ball, to calculate the attraction of the earth as well as we determine it by actually observing the motion of falling bodies around us.

Thus it is that when a planet, like Mars or Jupiter, has satellites revolving round it, astronomers on the earth can observe the attraction of the planet on its satellites and thus determine its mass. The rule for doing this is very simple. The cube of the distance between the planet and satellite is divided by the square of the time of revolution of the satellite. The quotient is a number which is proportional to the mass of the planet. The rule applies to the motion of the moon round the earth and of the planets round the sun. If we divide the cube of the earth's distance from the sun, say 93,000,000 miles, by the square of 365 ¼, the days in a year, we shall get a certain quotient. Let us call this number the sun-quotient. Then, if we divide the cube of the moon's distance from the earth by the square of its time of revolution, we shall get another quotient, which we may call the earth-quotient. The sun-quotient will come out about 330,000 times as large as the earth-quotient. Hence it is concluded that the mass of the sun is 330,000 times that of the earth; that it would take this number of earths to make a body as heavy as the sun.

I give this calculation to illustrate the principle; it must not be supposed that the astronomer proceeds exactly in this way and has only this simple calculation to make. In the case of the moon and earth, the motion and distance of the former vary in consequence of the attraction of the sun, so that their actual distance apart is a changing quantity. So what the astronomer actually does is to find the attraction of the earth by observing the length of a pendulum which beats seconds in various latitudes. Then, by very delicate mathematical processes, he can find with great exactness what would be the time of revolution of a small satellite at any given distance from the earth, and thus can get the earth-quotient.

But, as I have already pointed out, we must, in the case of the planets, find the quotient in question by means of the satellites; and it happens, fortunately, that the motions of these bodies are much less changed by the attraction of the sun than is the motion of the moon. Thus, when we make the computation for the outer satellite of Mars, we find the quotient to be $\frac{1}{3,093,500}$ that of the sun-quotient. Hence we conclude that the mass of Mars is $\frac{1}{3,093,500}$ that of the sun. By the corresponding quotient, the mass of Jupiter is found to be about $\frac{1}{1047}$ that of the sun, Saturn $\frac{1}{3500}$, Uranus $\frac{1}{22,700}$, Neptune $\frac{1}{19,500}$.

We have set forth only the great principle on which the astronomer has proceeded for the purpose in question. The law of gravitation is at the bottom of all his work. The effects of this law require mathematical processes which it has taken two hundred years to bring to their present state, and which are still far from perfect. The measurement of the distance of a satellite is not a job to be done in an evening; it requires patient labor extending through months and years, and then is not as exact as the astronomer would wish. He does the best he can, and must be satisfied with that.

CHAPTER NINE
The Sun

The chapters that follow are taken from Astronomy for Everybody, *whose subtitle—"A Popular Exposition of the Wonders of the Heavens"—has to my ears at least a wonderfully dawn-of-the-twentieth century ring to it. This was the heyday of the great expositions, or world's fairs, with two of them, the Chicago exposition of 1893 and the Pan-American Exposition held in Buffalo, N.Y., in 1901, still fresh in popular memory. So the conceit of providing a similar tour of the wonders of the universe between the pages of a book would have been a familiar one to Newcomb and his readers. Those parts of the book reprinted here provide a good sense of what we knew (and what we didn't) about the other members of the solar system at the start of the new century.*

I n a description of the solar system its great central body is naturally the first to claim our attention. We see that the sun is a shining globe. The first questions to present themselves to us are about the size and distance of this globe. It is easy to state its size when we know its distance. We know by measurement the angle subtended by the sun's diameter. If we draw two lines making this angle with each other, and continue them indefinitely through the celestial spaces, the diameter of the sun must be equal to the distance apart of the lines at the distance of the sun. The exact determination is a very simple problem of trigonometry. It will suffice at present to say that the measure of the apparent diameter of the sun, or the angle which it subtends to our eye, is 32 minutes, making this angle such that the distance of the sun is about 107.5 times its diameter in miles. If, then, we know the distance of the sun, we have only to divide it by 107.5 to get the sun's diameter.

The various methods of determining the distance of the sun will be described in our chapter stating how distances in the heavens are measured.[1] The result of all the determinations is that the distance is very nearly 93 million miles, perhaps one or two hundred thousand miles more. Taking the round number, and dividing by 107.5, we find the diameter to be about 865,000 miles. This is about 110 times the diameter of the earth. It follows that the volume or bulk of the sun is more than 1,300,000 times that of the earth.

The sun's importance to us arises from its being our great source of heat and light. Were these withdrawn, not only would the world be enveloped in unending night, but, in the course of a short time, in eternal frost. We all know that during a clear night the surface of the earth grows colder through the radiation into space of the heat received from the sun during the day. Without our daily supply, the loss of heat would go on until the cold around us would far exceed that which we now experience in the polar regions. Vegetation would be impossible. The oceans would freeze over, and all life on the earth would soon be extinct.

The surface of the sun, which is all we can see of it, is called the *photosphere*. This term is used to distinguish the visible surface from the vast invisible interior of the sun. To the naked eye, the photosphere looks entirely uniform. But through a telescope we see that the whole surface has a mottled appearance, which has been aptly compared to that of a plate of rice soup. Examination under the best conditions shows that this appearance is due to minute and very irregular grains which are scattered all over the photosphere.

When we carefully compare the brightness of different regions of the photosphere, we find that the apparent centre of the disk is brighter than the edge. The difference can be seen even without a telescope, if we look at the sun through a dark glass, or when it is setting in a dense haze. The falling off in the light is especially rapid as we approach the extreme edge of the disk, where it is little more than half as bright as at the centre. There is also a difference of colour, the light of the edge having a lurid appearance as compared with that of the center.

All this shows that the light of the sun is absorbed by an atmosphere surrounding the sun. We readily see that, the sun being a globe, the light which we receive from the edge of its disk leaves it obliquely, while that from the centre leaves it perpendicularly. The more obliquely the light

---

[1] That particular chapter of *Astronomy for Everybody* is not reprinted in this collection, but the principle is similar to that employed in determining the distances of the nearer stars by means of their parallax, as discussed on page 12.

comes from the surface, the greater the thickness of the sun's atmosphere through which it must pass, and hence the greater the portion lost by the absorption of that atmosphere. The sun's atmosphere, like our own, absorbs the green and blue rays more than the red. For this reason the light has a redder tint when it comes from near the edge of the disk.

ROTATION OF THE SUN

Careful observations show that the sun, like the planets, rotates on an axis passing through its center. Using the same terms as in the case of the earth, we call the points in which the axis intersects the surface the *poles* of the sun, and the circle around it halfway between the poles the sun's *equator*. The period of rotation is about 26 days. As the distance around the sun is more than 110 times that round the earth, the speed of rotation must be more than four times that of the earth's rotation to make it complete the circuit in the time that it does. At the sun's equator the speed is more than a mile a second.

The most curious feature of this rotation is that it is completed in less time at the equator than at a distance on each side of the equator. Were the sun a solid body, like the earth, all its parts would have to rotate at the same time. Hence the sun is not a solid body, but must be either liquid or gaseous, at least at its surface.

The equator of the sun is inclined six degrees to the plane of the earth's orbit. Its direction is such that in our spring months the north pole is turned six degrees away from us and the central point of the apparent disk is about that amount south of the sun's equator. In our summer and autumn months this is reversed.

THE SUN'S DENSITY AND GRAVITY

By the mean density of the sun we refer to the average specific gravity of the matter composing it, or the ratio of its weight to that of an equal volume of water. It is known that the density is only about one-fourth that of the earth, and about four-tenths greater than that of water. Stated with more exactness, the figures are:

Density of sun: Density of earth = 0.2554.
Density of sun: Density of water = 1.4115.

The mass or weight of the sun is about 334,000 times that of the earth.

The force of gravity at the sun's surface is 27 times that of the earth. If it were possible for a human being to be placed there, an ordinary man would weigh two tons, and be crushed by his own weight.

SPOTS ON THE SUN

When the sun is carefully examined with a telescope, one or more seemingly dark spots will generally, though not always, be seen on its surface. These are, of course, carried around by the rotation of the sun, and it is by means of them that the time of rotation is most easily determined. If a spot appears at the center of the disk it will, in six days, be carried to the western edge, and there disappear. At the end of about two weeks it will reappear at the eastern edge unless it has, in the meantime, died away, which is frequently the case.

The spots have a wide range in size. Some are very minute points, barely visible in a good telescope, while on rare occasions one is large enough to be seen with the naked eye through a dark glass. They frequently appear in groups, and a group may sometimes be made out with the naked eye as a minute patch when the individual spots cannot be seen.

When the air is steady, and a good-sized spot is carefully examined with a telescope, it will be seen to be composed of a dark central region or nucleus, surrounded by a shaded border. If all the conditions are favorable, this border will appear striated, like the edge of a thatched roof. The appearance is represented in the cut, which also shows the mottling of the photosphere.

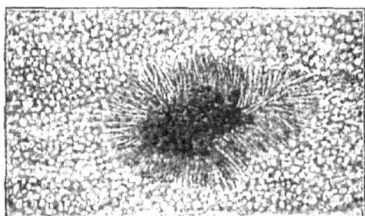

Appearance of a sunspot under high magnifying power, showing also the mottling of the photosphere.

The spots are of the most varied and irregular forms, frequently broken up in many ways. The shaded border or the thatched lines which form it frequently encroaches on the nucleus or may, in places, extend quite across it. A most remarkable law connected with the spots, which has been established by nearly three centuries of observation, is that their frequency varies in a regular period of eleven years and about forty days. During a certain year no spots will be visible for about half of the time. This was the case in 1889 and again in 1900. The year following a slightly greater number will show themselves; and they will increase year after year for about five years. Then the frequency will begin to diminish, year after year, until the cycle is completed, when it will again begin to increase. These mutations have been traced back to the time of Galileo, although it was not till about 1825 that they were found by Schwabe to take place in a regular period. . . .

Another noteworthy law connected with the sun's spots is that they are not found all over the sun; but only in certain regions of solar latitude. They are rather rare on the sun's equator, but become more frequent as we go north or south of the equator till we get to fifteen degrees of latitude, north or south. From this region to twenty degrees the frequency is greatest; then it falls off, so that beyond thirty degrees a spot is rarely seen. These regions are shown in the accompanying figure, where the shading is darker the more frequent the spots. If we made a white globe to represent the sun, and made a black dot on it for every spot during a number of years, the dotting would make the globe look as represented in the figure.

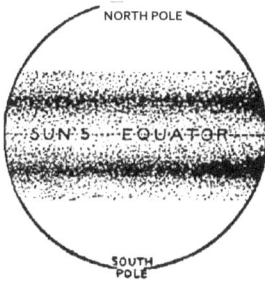

Frequency of sunspots in different latitudes on the sun.

THE FACULAE

Collections of numerous small spots brighter than the photosphere in general are frequently seen on the sun. These are often seen in the neighborhood of a sunspot, and occur most frequently in the regions of greater spot frequency, but are not entirely confined to those regions. They are, however, rare near the poles of the sun.

That the spots and faculae proceed from some one general cause has been brought out by the spectro-heliograph, an instrument devised by Professor George E. Hale for taking photographs of the sun by the light of a single ray of the spectrum—that emitted by calcium, for example. The effect is the same as if we should look at the sun through a glass which would allow the rays of calcium vapor to pass, but would absorb all the others. We should then see the calcium light of the sun and no other.

When the sun is photographed by calcium light with this instrument, the result is wonderful. The sunspot regions are now seen to be brighter than the others, and faculae are found on every part of the sun. We thus learn that eruptions of gas, of which calcium is the best marked ingredient, are taking place all the time; but they are more numerous in the sunspot zones than elsewhere. The sunspots are therefore the effect of operations going on all the time, all over the sun, but giving rise to a spot only in the exceptional cases when they are very intense.

It was formerly supposed that the spots were openings or depressions in the photosphere, showing a darker region within. This view was based on the belief that, when a spot was near the edge of the sun's disk, the shaded border next the edge looked broader than the other. But this view is now abandoned. We cannot certainly say that a spot is either above or below the photosphere. We shall hereafter see that the latter is not a mere surface as it seems to us, but a shell or covering many miles, perhaps a hundred or more, in thickness. The spots doubtless belong to this shell, being cooler portions of it, but lying neither above nor below it.

THE PROMINENCES AND CHROMOSPHERE

The next remarkable feature of the sun to be described consists in the prominences. Our knowledge of these objects has an interesting history—which will be mentioned in describing eclipses of the sun.[1] The spectroscope shows us that large masses of incandescent vapour burst forth from every part of the sun. They are of such extent that the earth, if immersed in them, would be as a grain of sand in the flame of a candle. They are thrown up with enormous velocity, sometimes hundreds of miles a second. Like the faculae, they are more numerous in the sunspot zones, but are not confined to those zones. The glare around the sun caused by the reflection of light by the air renders them entirely invisible to vision, even with the telescope, except when, during total eclipses of the sun, the glare is cut off by the intervention of the moon. They may then be seen, even with the naked eye, rising up as if from the black disk of the moon.

The prominences seem to be of two forms, the eruptive and the cloud-like. The first rise from the sun like immense sheets of flame; the latter seem to be at rest above it, like clouds floating in the air. But there is no air around the sun for these objects to float in, and we cannot certainly say what supports them. Very likely, however, it is a repulsive force of the sun's rays.

Spectrum analysis shows that these prominences are composed mostly

[1] See Chapter 7 in this collection.

of hydrogen gas, mixed with the vapors of calcium and magnesium. It is to the hydrogen that they owe their red colour. Continued study of the prominences shows them to be connected with a thin layer of gases which surrounds and rests upon the photosphere. This layer is called the chromosphere, from its deep red colour, similar to that of the prominences. As in the case of the latter, most of its light seems to be that of hydrogen; but it contains many other substances in seemingly varying proportions.

The last appendage of the sun to be considered is the corona. This is seen only during total eclipses as a soft effulgence surrounding the sun, and extending from it in long rays, sometimes exceeding the diameter of the sun in length. Its exact nature is still in doubt.

HOW THE SUN IS MADE UP

Let us now recapitulate what makes up the sun as we see and know it.

We have first the vast interior of the globe which, of course, we can never see.

What we see when we look at the sun is the shining surface of this globe, the photosphere. It is not a real surface, but more likely a gaseous layer several hundred miles deep which we cannot distinguish from a surface. This layer is variegated by spots, and in or over it rise the faculae.

On the top of the photosphere rests the layer of gases called the chromosphere, which can be observed at any time with a powerful spectroscope, but can be seen by direct vision only during total eclipses.

Through or from the red chromosphere are thrown up the equally red flames called the prominences.

Surrounding the whole is the corona.

Such is the sun as we see it. What can we say about what it really is? First, is it solid, liquid, or gaseous?

That it is not solid we have already shown by the law of rotation. It cannot be a liquid like molten metal, because it sends off from its surface such a flood of heat as would cool off and solidify molten metal in a very short time. For more than thirty years [i.e., since the 1860s] it has been understood that the interior of the sun must be a mass of gas, compressed to the density of a liquid by the enormous pressure of its superincumbent portions. But it was still supposed that the photosphere might be in the nature of a crust and the whole sun like an immense bubble. This view, however, seems no longer tenable. It does not seem likely that there is any solid matter on the sun.

Attempts have sometimes been made to learn the temperature of the photosphere. It probably exceeds any that we can produce on earth, even that of the electric furnace, else how could calcium, the metallic base of lime, one of the most refractory of substances, exist there in a state of vapor? We all know that the air around us becomes cooler and rarer as we ascend above the surface of the earth, owing to the action of gravity and the consequent weight of the atmosphere, which gives rise to a constantly increasing pressure as we descend. Now, gravity at the sun is 27 times as powerful as on the earth. Hence, going downward, temperature and pressure increase at a far more rapid rate on the sun than on the earth. Even in the photosphere the temperature is such that "the elements melt with fervent heat." And, as we go below the surface, the heat must increase by hundreds of degrees for every mile that we descend. The result is that in the interior the gases of the sun are subjected to two opposing forces which grow more and more intense. These are the expansive force of the heat and the compressing force of the gases above, produced by the enormous force of gravity of the sun.

The forces thus set in play merely in the outer portions of the sun's globe are simply inconceivable. Perhaps the explosion of the powder when a thirteen-inch cannon is fired is as striking an example of the force of ignited gases as we are familiar with. Now suppose every foot of space in a whole county covered with such cannon, all pointed upward and all being discharged at once. The result would compare with what is going on inside the photosphere about as a boy's popgun compares with the cannon.

THE SOURCE OF THE SUN'S HEAT

Perhaps, from a practical point of view, the most comprehensive and important problem of science is: How is the sun's heat kept up? Before the laws of heat were fully apprehended this question was not supposed to offer any difficulties. Even to this day it is supposed by those not acquainted with the subject, that the heat which we receive from the sun may arise in some way from the passage of its rays through our atmosphere, and that, as a matter of fact, the sun may not radiate any actual heat at all—may not be an extremely hot body. But modern science shows that heat cannot be produced except by the expenditure of some form of energy. The energy of the sun is necessarily limited in quantity and is continually being lost through radiation.

It is very easy to imagine the sun as being something like a white-hot cannon ball, which is cooling off by sending its heat in all directions, as such a ball does. We know by actual observation how much heat the sun sends to us. It may be expressed in the following way:

Imagine a shallow basin with a flat bottom, and a depth of one centimetre, that is, about four-tenths of an inch. Let the basin be filled with water, the latter then being one centimetre deep. Expose such a basin to the rays of the vertical sun. The heat which the sun will radiate to them will be sufficient to warm the water about three-and-a-half or four degrees Centigrade, or not very far from seven degrees Fahrenheit, in one minute. It follows that if we suppose a thin spherical shell of water, one centimeter thick, of the same radius as the earth's orbit, and having the sun in its centre, that shell of water will be heated with the rapidity just mentioned. The heat which it receives will be the total amount radiated by the sun. We can thus define how much heat the sun loses every minute, day, and year.

A very simple calculation will show that if the sun were of the nature of a white-hot ball it would cool off so rapidly that its heat could not last more than a few centuries. But it has in all probability lasted millions of years. Whence, then, comes the supply? The answer of modern science to this question is that the heat radiated from the sun is supplied by the contraction of size as heat is lost. We all know that in many cases when motion is destroyed heat is produced. When a cannon shot is fired at the armour plate of a ship of war, the mere stroke of the shot makes both plate and shot hot. The blacksmith can make iron hot by hammering it.

These facts have been generalized into the statement that whenever a body falls and is stopped in its fall by friction, or by a stroke of any sort, heat is produced. From the law governing the case, we know that the water of Niagara, after it strikes the bottom of the falls, must be about one quarter of a degree warmer than it was during the fall. We also know that a hot body contracts in volume when cooled. The contraction of a gaseous body, such as we believe the sun to be, is greater than that of a solid or liquid. The heat of the sun is radiated from streams of matter constantly rising from the interior, which radiate their heat when they reach the surface. Being cooled they fall back again, and the heat caused by this fall is what keeps the sun hot.

It may seem almost impossible that heat sufficient to last for millions of years could be generated in this way; but the known force of gravity at the surface of the sun enables us to make exact computations on the subject. It is thus found that in order to keep up the supply of heat it is only necessary that the diameter of the sun should contract about a mile in twenty-five years—or four miles in a century. This amount would not be perceptible until after thousands of years. Yet the process of contraction must come to an end some time. Therefore, if this view is correct, the life of the sun must have a limit. What its limit may be we cannot say with exactness, we only know that it is several millions of years, but not many millions.

The same theory implies that the sun was larger in former times than it is now, and must have been larger and larger every year that we go back into its history. There was a time when it must have been as large as the whole solar system. In this case it could have been nothing but a nebula. We thus have the theory that the sun and solar system have resulted from the contraction of a nebula—through millions of years. This view is familiarly known as the nebular hypothesis.

The question whether the nebular hypothesis is to be accepted as a proved result of science is one on which opinions differ. There are many facts which support it—such as the interior heat of the earth and the revolution and rotation of the planets all in the same direction. But cautious and conservative minds will want some further proof of the theory before they regard it as absolutely established. Even if we accept it, we still have open the question: How did the nebula itself originate, and how did it begin to contract? This brings us to the boundary where science can propound a question but cannot answer it.

CHAPTER TEN

The Earth

The globe on which we live, being one of the planets, would be entitled to a place among the heavenly bodies even if it had no other claims on our attention. Insignificant though it is in size when compared with the great bodies of the universe, or even with the four giant planets of our system, it is the largest of the group to which it belongs. Of the rank which it might claim as the abode of man we need not speak.

What is the earth? We may describe it in the most comprehensive way as a globe of matter nearly 8000 miles in diameter, bound together by the mutual gravitation of its parts. We all know that it is not exactly spherical, but bulges out very slightly at the equator. The problem of determining its exact shape and size is an extremely difficult one, and we cannot say that an entirely satisfactory result is yet reached. The difficulty is obvious enough. There is no way of measuring distances across the great oceans. The measurements are necessarily limited to such islands as are visible from the coasts of the continents or from each other. Of course, the measures cannot be extended to either pole. The size and shape must therefore be inferred from the measures across or along the continents. Owing to the importance of such work, the leading nations have from time to time entered into it. Quite recently our Coast and Geodetic Survey has completed the measurement of a line of triangles extending from the Atlantic to the Pacific Oceans. North and south measurements both on the Atlantic and Pacific coasts have been executed or are in progress. The English have from time to time made measures of the same sort in Africa, and the Russians and Germans on their respective territories. Nearly all these measures are now being combined in a work carried on by the International Geodetic Association, of which the geodetic authorities of the principal countries are members.

The latest conclusions on the subject may be summed up thus. We remark in the first place that by the figure of the earth geodetists do not mean the figure of the continents, but of the ocean level as it would be if canals admitting the water of the oceans were dug through the continents. The earth thus defined is approximately an ellipsoid, of which the smaller diameter is that through the poles, and which has about the following dimensions:

Polar diameter, 7,899.6 miles, or 12,713.0 kilometres.
Equatorial diameter, 7,926.6 miles, or 12,756.5 kilometres.

It will be seen that the equatorial diameter is twenty-seven miles or forty-three kilometres greater than the polar.

THE EARTH'S INTERIOR

What we know of the earth by direct observation is confined almost entirely to its surface. The greatest depth to which man has ever been able to penetrate compares with the size of the globe only as the skin of an apple does to the body of the fruit itself.

I shall first invite the reader's attention to some facts about weight, pressure, and gravity in the earth. Let us consider a cubic foot of soil forming part of the outer surface of the earth. This upper cubic foot presses upon its bottom with its own weight, perhaps 150 pounds. The cubic foot below it weighs an equal amount, and therefore presses on its bottom with a force equal to its own weight with the weight of the other foot added to it. This continual increase of pressure goes on as we descend. Every square foot in the earth's interior sustains a pressure equal to the weight of a column of the earth a foot square extending to the surface. Not many yards below the surface this pressure will be measured in tons; at the depth of a mile it may be thirty or forty tons; at the depth of one hundred miles, thousands of tons; continually increasing to the center. Under this enormous pressure the matter composing the inner portion of the earth is compressed to the density of a metal. By a process which we will hereafter describe, the mean density of the earth is known to be 5 ½ times that of water, while the superficial density is only two or three times that of water.

One of the most remarkable facts about the earth is that the temperature continually increases as we penetrate below the surface in deep mines. The rate of increase is different in different latitudes and regions. The general average is one degree Fahrenheit in fifty or sixty feet.

The first question to suggest itself is, how far toward the earth's center does this increase of temperature extend? The most that we can say is that it cannot be merely superficial, because, in that case, the exterior portions would have cooled off long ago, so that we should have no considerable increase of heat as we went down. The fact that the heat has been kept up during the whole of the earth's existence shows that it must still be very intense toward the centre, and that the rate of increase near the surface must go on for many miles into the interior.

At this rate the material of the earth would be red hot at a depth of ten or fifteen miles, while at one or two hundred miles the heat would be sufficient to melt all the substances which form the earth's crust. This fact suggested to geologists the idea that our globe is really a molten mass, like a mass of melted iron, covered by a cool crust a few miles thick, on which we dwell. The existence of volcanoes and the occurrence of earthquakes gave additional weight to this view, as did also other geological evidence, showing changes in the earth's surface which appeared to be the result of a liquid interior.

But in recent years the astronomer and physicist have collected evidence, which is as conclusive as such evidence can be, that the earth is solid from center to surface, and even more rigid than a similar mass of steel. The subject was first developed most fully by Lord Kelvin, who showed that, if the earth were a fluid, surrounded by a crust, the action of the moon would not cause tides in the ocean, but would merely tend to stretch out the entire earth in the direction of the moon, leaving the relative positions of the crust and the water unchanged.

Equally conclusive is the curious phenomenon which we shall describe presently of the variation of latitudes on the earth's surface. Not only a globe of which the interior is soft, but even a globe no more rigid than steel could not rotate as the earth does.

How, then, are we to reconcile the enormous temperature and the solidity? There seems to be only one solution possible. The matter of the interior of the earth is kept solid by the enormous pressure. It is found experimentally that when masses of matter like the rocks of the earth are raised to the melting point, and then subjected to heavy pressure, the effect of the pressure is to make them solid again. Thus, as we increase the temperature we have only to increase the pressure also to keep the material of the earth solid. And thus it is that, as we descend into the earth, the increase of pressure more than keeps pace with the rise of temperature, and thus keeps the whole mass solid.

GRAVITY AND DENSITY OF THE EARTH

Another interesting question connected with the earth is that of its density, or specific gravity. We all know that a lump of lead is heavier than an equal lump of iron, and the latter heavier than an equal lump of wood. Is there any way of determining what a cubic foot of earth would weigh if taken out from a great depth of its vast interior? If there is, then we can determine what the actual weight of the whole earth is. The solution depends on the gravitation of matter.

Every child is familiar with gravitation from the time it begins to walk, but the profoundest philosopher knows nothing of its cause, and science has not discovered anything respecting it except a few general facts. The widest and most general of these facts, which may be said to include the whole subject, is Sir Isaac Newton's theory of gravitation. According to this theory, the mysterious force by which all bodies on the surface of the earth tend to fall toward its center does not reside merely in the centre of the earth, but is due to an attraction exerted by every particle of matter composing our globe. Whether this was the case was at first an open question. Even so great a philosopher and physicist as Huyghens believed that the power resided in the earth's centre, and not in every particle, as Newton supposed. But the latter extended his theory yet farther by showing that every particle of matter in the universe, so far as we have yet ascertained, attracts every other particle with a force that diminishes as the square of the distance increases. This means that at twice the distance the attraction will be divided by four; at three times by nine; at four times by sixteen, and so on.

Granting this, it follows that all objects around us have their own gravitating power, and the question arises: Can we show this power by experiment, and measure its amount? The mathematical theory shows that globes should attract small bodies at their surfaces with a force proportioned to their diameter. A globe two feet in diameter, of the same specific gravity as the earth, should attract with a force one twenty-millionth of the earth's gravity.

In recent times several physicists have succeeded in measuring the attraction of globes of lead having a diameter of a foot, more or less. This measurement is the most delicate and difficult that has ever been made, and the accuracy which seems to have been reached would have been incredible a few years ago. The apparatus used is, in its principle, of the simplest kind. A very light horizontal rod is suspended at its centre by a thread of the finest and most flexible material that can be obtained. This rod is balanced by having a small ball attached to each end. What is measured is the attraction of the globes of lead upon these two balls. The former are placed in such a position as to unite their attraction in giving the rod a slight twisting motion in the horizontal plane. To appreciate the difficulties of the case, we must call to mind that the attraction may not amount to the ten-millionth part of the weight of the little balls. It would be difficult to find any object so light that its weight would not exceed this force. To compare the weight of a fly with it would be like comparing the weight of an ox with that of a dose of medicine. Not only the weight of a mosquito but even of its finest limb might exceed the quantity to be measured. If a mosquito were placed under a microscope an expert operator could cut off from one antenna a piece small enough to express the force measured.

Yet the determination of this force has been made with such precision that the results of the two latest investigators do not differ by a thousandth part. These were Professor Boys, F.R.S., of Oxford, England, and Dr. Karl Braun, S.J., of Marienschein, in Bohemia. They worked independently at the problem, meeting and overcoming innumerable difficulties one after another, getting greater and greater delicacy and precision in their apparatus, and finally published their results almost at the same time, the one in England, the other in Austria. The outcome of their experiments is that the mean density of the earth is slightly more than 5 ½ times that of water. This is a little less than the density of iron, but much more than that of any ordinary stone. As the mean density of the materials which compose the earth's crust is scarcely more than one half of this amount, it follows that near the center the matter composing the earth must be compressed to a density not only far exceeding that of iron, but probably that of lead.

The attraction of mountains has been known for more than a hundred years. It was first demonstrated by Maskelyne about 1775 in the case of Mount Schehallion, in Scotland. In all mountain regions where very accurate surveys are made the attraction of mountains upon the plumb line is very evident.

VARIATIONS OF LATITUDE

We know that the earth rotates on an axis passing through the centre and intersecting the earth's surface at either pole. If we imagine ourselves standing exactly on a pole of the earth, with a flagstaff fastened in the ground, we should be carried round the flagstaff by the earth's rotation once in twenty-four hours. We should become aware of the motion by seeing the sun and stars apparently moving in the opposite direction in horizontal circles by virtue of the diurnal motion. Now, the great discovery of the variation of latitude is this: The point in which the axis of rotation intersects the surface is not fixed, but moves around in a somewhat variable and irregular curve, contained within a circle nearly sixty feet in diameter. That is to say, if standing at the north pole we should observe its position day by day, we should find it moving one, two, or three inches every day, describing in the course of time a curve around one central point, from which it would sometimes be farther away and sometimes nearer. It would make a complete revolution in this irregular way in about fourteen months.

Since we have never been at the pole, the question may arise: How is this known? The answer is that by astronomical observations we can, on any night, determine the exact angle between the plumb line at the place where we stand and the axis on which the earth is rotating on that particular day. Four or five stations for making these observations were established around the earth in 1900 by the International Geodetic Association. One of these stations is near Gaithersburg, Maryland, another is on the Pacific coast, a third is in Japan, and a fourth in Italy. Before these were established, observations having the same object were made in various parts of Europe and America. The two most important stations in the latter region were those of Professor Rees of Columbia University, New York, and of Professor Doolittle, first at Lehigh, and later at the Flower Observatory, near Philadelphia.

The variation which we have described was originally demonstrated by S. C. Chandler, of Cambridge, in 1890 by means of a great mass of astronomical observations not made for this special purpose. Since then investigation has been going on with the view of determining the exact curve described. What has been shown thus far is that the variation is much wider some years than others, being quite considerable in 1891, and very small in 1894. It appears that in the course of seven years there will be one in which the pole describes the greater part of a comparatively wide circle, while three or four years later it will for several months scarcely move from its central position.

If the earth were composed of a fluid, or even of a substance which would bend no more than the hardest steel, such a motion of the axis as this would be impossible. Our globe must therefore, in the general average, be more rigid than steel.

THE ATMOSPHERE

The atmosphere is astronomically, as well as physically, a most important appendage of the earth. Necessary though it is to our life it constitutes one of the greatest obstructions with which the astronomer has to deal. It absorbs more or less of all the light that passes through it, and thus slightly changes the colour of the heavenly objects as we see them, and renders them somewhat dimmer, even in the clearest sky. It also refracts the light passing through it, causing it to describe a slightly curved line, concave toward the earth, instead of passing straight to the astronomer's eye. The result of this is that the stars appear slightly higher above the horizon than they actually are. The light coming directly down from a star in the zenith suffers no refraction. The latter increases as the star is farther from the zenith, but even forty-five degrees away it is only one minute of arc, about the smallest amount that the unaided eye can plainly perceive; yet this is a very important quantity to the astronomer. The nearer the object is to the horizon the greater the rate at which the refraction increases; twenty-eight degrees above the horizon it is about twice as great as at forty-five degrees; at the horizon it is more than one half a degree, that is more than the whole diameter of the sun or moon. The result is that when we see the sun just about to touch the horizon at sunset or sunrise its whole body is in reality below the horizon. We see it only in consequence of the refraction of its light. Another result of the rapid increase near the horizon is that, in this position, the sun looks decidedly flattened to the eye, its vertical diameter being shorter than the horizontal one. Anyone may notice this who has an opportunity to look at the sun as it is setting in the ocean. It arises from the fact that the lower edge of the sun is refracted more than the upper edge.

When the sun sets in the ocean in the clear air of the tropics a beautiful effect may be noticed, which can rarely or never be seen in the thicker air of our latitudes. It arises from the unequal refraction of the rays of light by the atmosphere. Like a prism of glass the atmosphere refracts the red rays the least and the successive spectral colours, yellow, green, blue, and violet, more and more. The result is that, as the edge of the sun is disappearing in the ocean, these successive rays are lost sight of in the

79

same order. Two or three seconds before the sun has disappeared, the little spark of its limb which still remains visible is seen to change colour and rapidly grow paler. This tint changes to green and blue, and finally the last glimpse which we see is that of a disappearing flash of blue or violet light.

CHAPTER ELEVEN
The Moon

We know far more about the moon than any of our other neighbors in space, simply because it is so close: less than a quarter of a million miles away. Even in 1902, Newcomb was able to set down a basic picture of the moon that is still in its broadest outlines accurate today. Along the way, he tackled two issues that still bedevil popular science writers (and their readers) more than a century later: how can the moon present only one face to the earth, yet still rotate on its axis; and why are there two high tides each day, not just one?

About one hundred years ago there was an unpopular professor in the Government Polytechnique School of Paris, still the great school of mathematics for the French public service, who loved to get his students into difficulties. One morning he addressed one of them the question: "Monsieur, have you ever seen the moon?"

"No, sir," replied the student, suspecting a trap.

The professor was nonplussed. "Gentlemen," said he, "see Mr. ——, who professes never to have seen the moon!"

The class all smiled.

"I admit that I have heard it spoken of," said the student, "but I have never seen it."

I take it for granted that the reader has been more observant than the French student professed to be, and that he has not only seen the moon, but knows the phases through which it goes and is familiar with the fact that it describes a monthly course around the earth. I also suppose that he knows the moon to be a globe, although, to the naked eye, it seems like a flat disk. The globular form is, however, very evident when we look at it with a small telescope.

Various methods and systems of measurement all agree in placing the moon at an average distance of a little less than 240,000 miles. This distance is obtained by direct measure of the parallax, as will be explained hereafter, and also by calculating how far off the moon must be in order that, being projected into space, it may describe an orbit around the earth in the time that it actually does perform its round. The orbit is elliptic, so that the actual distance varies. Sometimes it is ten or fifteen thousand miles less, at other times as much more, than the average.

The diameter of the moon's globe is a little more than one-fourth that of the earth; more exactly, it is 2160 miles. The most careful measures show no deviation from the globular form except that the surface is very irregular.

REVOLUTION AND PHASES OF THE MOON

The moon accompanies the earth in its revolution round the sun. To some the combination of the two motions seems a little complex; but it need not offer any real difficulty. Imagine a chair standing in the center of a railway car in rapid motion, while a person is walking around it at a distance of three feet. He can go round and round without varying his distance from the chair and without any difficulty arising from the motion of the car.

Thus the earth moves forward in its orbit, and the moon continually revolves around it without greatly varying its distance from us.

The actual time of the moon's revolution around the earth is twenty-seven days, eight hours; but the time from one new moon to another is twenty-nine days, thirteen hours. The difference arises from the earth's motion around the sun; or, which amounts to the same thing, the apparent motion of the sun along the ecliptic. To show this, let AC be a small arc of the earth's orbit around the sun. Suppose that at a certain time the earth is at the point E, and the moon at the point M, between the earth and the sun. At the end of twenty-seven days, eight hours the earth will have moved from E to F. While the earth is making this mo-

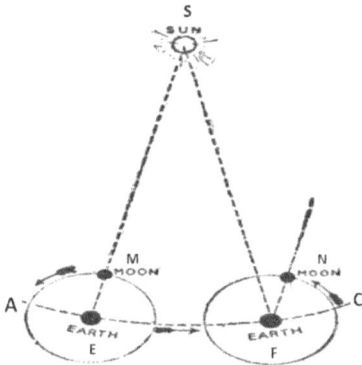

Revolution of the moon round the earth

tion the moon will have moved around the orbit in the direction of the arrows, so as to have reached the point N. At the moment when the lines EM and FN are parallel to each other, the moon will have completed her actual revolution, and will seem to be in the same place among the stars as before. But the sun is now in the direction FS. The moon therefore has to continue its motion before it catches up to the sun. This requires a little more than two days, and makes the whole time between two new moons 29½ days.

The varying phases of the moon depend upon its position with respect to the sun. Being an opaque globe, without light of its own, we see it only as the light of the sun illuminates it. When it is between us and the sun its dark hemisphere is turned toward us, and it is entirely invisible. The time of this position in the almanacs is called "new moon," but we cannot commonly see the moon for nearly two days after this time, because it is lost in the bright twilight of evening. On the second and third day, however, we see a small portion of the illuminated globe, having the familiar form of a thin crescent. This crescent we commonly call the new moon, although the time given in the almanac is several days earlier.

In this position, and for several days longer, we may, if the sky is clear, see the entire face of the moon, the dark parts shining with a faint gray light. This light is that which is reflected from the earth to the moon. An inhabitant of the moon, if there were such, would then see the earth in the sky like a full moon, looking much larger than the moon looks to us. As the moon advances in its orbit day after day, this light diminishes, and about the time of first quarter disappears from our sight owing to the brightness of the illuminated portion of the moon.

Seven or eight days after the almanac time of new moon, the moon reaches its first quarter. We then see half of the illuminated disk. During the week following, the moon has the form called gibbous. At the end of the second week the moon is opposite the sun, and we see its entire hemisphere like a round disk. This we call full moon. During the remainder of its course the phases recur in reverse order, as we all know.

We might regard all these recurrences as too well known to need description, yet, in *The Ancient Mariner*, a star is described as seen between the two horns of the moon as though there were no dark body there to intercept our view of the star. Probably more than one poet has described the new moon as seen in the eastern sky, or the evening full moon as seen in the west.

THE SURFACE OF THE MOON

We can see with the naked eye that the moon's surface is variegated by bright and dark regions. The latter are sometimes conceived to have a vague resemblance to the human face, the nose and eyes being especially prominent. Hence the "man in the moon." Through even the smallest telescopes we see that the surface has an immense variety of detail: and the more powerful the telescope the more details we see. The first thing to strike us on a telescopic examination will be the elevations, or mountains as they are commonly called. These are best seen about the time of the first quarter, because they then cast shadows. At full moon they cannot be so well made out, because we are looking straight down and see everything illuminated. Although these elevations and depressions are called mountains they are different in form from the ordinary mountains of the earth. There is, however, an almost exact resemblance between them and the craters of our great volcanoes. A very common form is that of a circular fort, one or more miles in diameter, with walls which may be thousands of feet high. The inside of this fort may be saucer shaped, a large portion of the surface being flat. At first quarter we can see the shadow of the walls cast upon the interior flat surface. In the centre a little cone is frequently seen. The interior surface is by no means perfectly flat and smooth. The higher power the more details we shall see. Just what these consist of it is impossible to say; they may be solid rock or they may be piles of loose stone. As we can see no object on the moon, even with the most powerful telescope, unless it is more than a hundred feet in diameter, we cannot say what the exact nature of the surface is in its minutest portions.

The early observers with the telescope supposed that the dark portions were seas and the brighter portions continents. This notion was founded on the fact that the darker portions looked smoother than the others. Names were therefore given to these supposed oceans, such as *Mare Procellarum*, the Sea of Storms; *Mare Serenitatis*, the Sea of Calms, etc. These names, fanciful though they be, are still retained to designate the large dark regions on the moon. A very slight improvement in the telescope, however, showed that the idea of these dark regions being oceans was an illusion. They are all covered with inequalities, proving that they must be composed of solid matter. The difference of aspect arises from the lighter or darker shade of the materials which compose the lunar surface. These are distributed over the surface of the moon in a very curious way. One of the most remarkable features is the long bright lines which radiate from certain points on the moon. A very low telescopic power will show the most remarkable of these; a good eye might even perceive it without a telescope. On the southern part of the moon's hemisphere, as we see it, is a large spot or region known as Tycho, and from this radiate a number of these bright streaks. The appearance is as if the moon had been cracked and the cracks filled up with melted white matter.

Whether we accept this view or not, it is impossible to examine the surface of the moon without the conviction that in some former age it was the seat of great volcanic activity. In the centre of all the great circular mountains we have described are craters which, it would seem, must have been those of volcanoes. Indeed, a hundred years ago it was supposed by Sir William Herschel that there was an active volcano on the moon, but it is now known that this appearance is due to the light of the earth reflected from a very bright spot on the moon's surface. It can be easily seen about the time of the new moon with a telescope of moderate size.

IS THERE AIR OR WATER ON THE MOON?

One of the most important questions connected with the moon is whether there is any air or water on its surface. To these the answer of science up to the present time is in the negative. Of course this does not mean that there can absolutely not be a drop of moisture nor the smallest trace of an atmosphere on our satellite; all we can say is that if any atmosphere surrounds the moon it is so rare that we have never been able to get any evidence of its existence. If the latter had such an appendage of even one-hundredth of the density of the earth's atmosphere, its existence would be made known to us by refraction of the light from a star seen alongside the moon. But not the slightest trace of any such refraction can be discovered. If there is any such liquid as water, it must be concealed in invisible crevices, or diffused through the interior. Were there any large sheets of water in the equatorial regions they would reflect the light of the sun day by day, and would thus become clearly visible. The water would also evaporate and form more or less of an atmosphere of watery vapor.

All this seems to settle another important question; namely, that of the habitability of the moon. Life, in the form in which it exists on our earth, requires water at least for its support, and in all its higher forms air also. We can hardly conceive of a living thing made of mere sand or other dry matter such as forms the lunar surface. If we supposed animals to walk about on the moon, it is difficult to imagine what they could eat. Our general conclusion must be that there is no life on the moon subject to the laws which govern life on the surface of this earth.

The total absence of air and water results in a state of things on the moon such as we never experience on the earth. So far as can be ascertained by the most careful examination, not the slightest change ever takes place on its surface. A stone lying on the surface of the earth is continually attacked by the weather and in the course of years is gradually disintegrated or washed away by the wind and water. But there is no weather on the moon, and a stone lying on its surface might rest there for unknown ages undisturbed by any cause whatever. The lunar surface is heated up when the sun shines on it and it cools off when the sun has set. Except for these changes of temperature there is absolutely nothing going on over the whole surface of the moon, so far as we can see. A world which has no weather and on which nothing ever happens—such is the moon.

ROTATION OF THE MOON

The rotation of the moon on its axis is a subject on which some are frequently so perplexed that we shall explain it. Anyone who has carefully examined this body knows that it always presents the same face to us. This shows that it rotates on its axis in the same time that it revolves around the earth. An idea frequently entertained is that this shows that it does not rotate at all, and many chapters have been written on this subject. The whole difficulty arises from the different ideas which people have of motion. In physics we say that a body does not rotate when, if a rod were passed through it, that rod always maintained the same direction when the body moved about. Let us suppose such a rod passed through the moon; then, if the latter did not rotate on its axis, the rod would maintain its same direction while the moon, revolving around the earth, would appear at different points in its orbit as we see it in the figure above. A very little study of this figure will show that as the moon went around we should successively see every part of its surface in succession if it did not rotate on its axis.

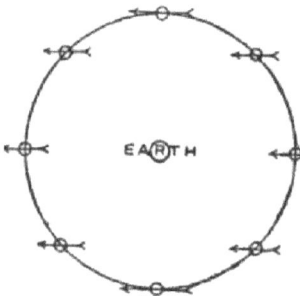

Showing how the moon would move if it did not rotate on its axis

HOW THE MOON PRODUCES THE TIDES

All of us who live on the seashore know that there is a rise and fall of the ocean which in the general average occurs about three-quarters of an hour later every day, and which keeps pace with the apparent diurnal motion of the moon. That is to say, if it is high tide today when the moon is in a certain position in the heavens, it will be high tide when the moon is in or near that position day after day, month after month, and year after year. We have all heard that the moon produces these tides by its attraction on the ocean. We readily understand that when the moon is above any region its attraction tends to raise the waters in that region; but the circumstance that most perplexes those who are not expert in the subject is that there are two tides a day, high tide occurring not only under the moon, but on the side of the earth opposite the moon. The explanation of this is that the moon really attracts the earth itself as well as it does the water. It continually draws the entire earth and everything upon it toward itself. As it goes round the earth in its monthly course, it thus keeps up a continual motion of the latter. If it attracted every part of the earth equally, the ocean included, there would then be no tides, and everything would go on on the earth's surface as if there were no attraction at all. But as the attraction is as the inverse square of the distance, the moon attracts the regions of the earth and oceans which are nearest to it more than the average, and those that are farthest from it less than the average.

To show the effect of these changes let A, C, and H be the three points on the earth attracted by the moon.

How the moon's pull on the earth and ocean produces two tides in a day

Since the moon attracts C more than A, it tends to pull C away from A and increase the distance between A and C. At the same time pulling H more than C it tends to increase the distance between H and C. If the whole earth was a fluid, the attraction of the moon would be simply to draw this fluid out into the form of an ellipsoid, of which the long diameter would be turned toward the moon. But the earth itself, being solid, cannot be drawn out into this shape, while the ocean, being fluid, is thus drawn out. The result is that we have high tides at the two ends of the ellipse into which the ocean is drawn, and low tides in the mid-region.

The complete explanation of the subject requires a statement of the laws of motion which cannot be made here. I will, however, remark that if the attraction of the moon on the earth were always in the same direction, the two bodies would be drawn together in a few days. But owing to the revolution of the moon round the earth the direction of the pull is always changing, so that the earth is, in the course of a month, only drawn about 3,000 miles from its mean position by the moon's pull.

It might be supposed that if the moon produces the tides in this way we should always have high tide when the moon is on the meridian and low tide when the moon is in the horizon. But such is not the case, for two reasons. In the first place it takes time for the moon to draw the waters out into the form of an ellipsoid, and when it once gives them the motion necessary to keep this form, that motion keeps up after the moon has passed the meridian, just as a stone continues to rise after it has left the hand or a wave goes forward by the momentum of the water. The other cause is found in the interruption of the motion by the great continents. The tidal wave, as it is called, meeting a continent, spreads out in one direction or the other, according to the lay of the land, and may be a long time in passing from one point to another. Thus arise all sorts of irregularities in the tides when we compare those in different places.

The sun produces a tide as well as the moon, but a smaller one. At the times of new and full moon the two bodies unite their forces and cause the highest and lowest tides. These are familiar to all dwellers on the seacoast and are called *spring tides*. About the time of the first and last quarters the attraction of the sun opposes that of the moon and the tides do not rise so high or fall so low, and these are called *neap tides*.

CHAPTER TWELVE
The Planet Mars

Though the state of astronomical knowledge has advanced enormously since Newcomb's time, one thing has not changed: no planet stirs more public interest than does Mars. When Astronomy for Everybody *was first published, Schiaparelli's initial reports of dark, thin lines on the disk of Mars—the Italian astronomer called them "channels," but this was mistranslated into English as "canals," thus implying an artificial origin—were less than two decades old, and the American astronomer Percival Lowell was in the process of publishing drawings of Mars that seemed to indicate the existence of a vast, intricate, globe-girdling network of canals. Lowell more than anyone else was responsible for the idea that Mars was inhabited by an advanced civilization struggling to survive on its dying, drying-out planet. Newcomb was skeptical about the canals' existence, with, as it turned out, good reason: they don't exist. His analysis of why certain observers saw them while others did not turned out to be essentially correct. Interestingly, later editions of* Astronomy for Everybody, *revised by Robert H. Baker, published after Newcomb's death, and incorporating subsequent research, took a much more sanguine view of the canals' existence. But Newcomb had got it right to begin with.*

M ore public interest has in recent years been concentrated on the planet Mars than on any other. Its resemblance to our earth, its supposed canals, oceans, climate, snowfall, etc., have all tended to interest us in its possible inhabitants. At the risk of disappointing those readers who would like to see certain proof that our neighboring world is peopled with rational beings, I shall endeavor to set forth what is actually known on the subject, distinguishing it from the great mass of illusion and baseless speculation which has crept into popular journals during the past twenty years.

We begin with some particulars which will be useful in recognizing the planet. Its period of revolution is 687 days, or forty-three days less than two years. If the period were exactly two years, it would make one revolution while the earth made two, and we should see the planet in opposition at regular intervals of two years. But, as it moves a little faster than this, it takes the earth from one to two months to catch up with it, so that the oppositions occur at intervals of two years and one or two months. This excess of one or two months makes up a whole year after eight oppositions; consequently, at the end of about seventeen years, Mars will again be in opposition at the same time of the year, and near the same point of its orbit, as before. In this period the earth will have made seventeen revolutions and Mars nine.

The difference of a month or so in the interval between oppositions is due to the great eccentricity of the orbit, which is larger than that of any other major planet except Mercury. Its value is 0.093, or nearly one-tenth. Hence, when in perihelion, it is nearly one-tenth nearer the sun than its mean distance, and when in aphelion nearly one-tenth farther. Its distance from the earth at opposition will be different by the same amount, measured in miles, and hence in a much larger proportion to the distance itself. If opposition occurs when the planet is near perihelion, the distance from earth is about 43 million miles; but if near the aphelion, about 60 million miles. The result of this is that, at a perihelion opposition, which can occur only in September, the planet will appear more than three times as bright as at an aphelion opposition, occurring in February or March. An opposition occurred near the end of March, 1903; the next following early in May, 1905. We shall then have oppositions near the end of June, 1907, and in August, 1909, which will be quite near to perihelion.

Mars, when near opposition, is easily recognised by its brilliancy, and by the reddish colour of its light, which is very different from that of most of the stars. It is curious that a telescopic view of the planet does not give so strong an impression of red light as does the naked eye view.

THE SURFACE AND ROTATION OF MARS

The great Huygens, who flourished between 1650 and 1700, studying Mars with the telescope, was the first one to recognize the variegated character of its surface, and to make a drawing of the appearance which it presented.

The features delineated by Huygens can be recognized and identified to this day. By watching them it was easy to see that the planet rotated on its axis in a little more than one of our days (24h. 37m.).

This time of rotation is the only definite and certain one among all the planets besides the earth. For 200 years Mars has rotated at exactly this rate, and there is no reason to suppose that the time will change appreciably any more than the length of our day will. The close approach to one of our days, the excess being only thirty-seven minutes, leads to the result that, on successive nights, Mars will, at the same hour, present nearly the same face to the earth. But, owing to the excess in question, it will always be a little farther behind on any one night than on the night before, so that, at the end of forty days, we shall have seen every part of the planet that is presented to the earth.

All that was known of Mars up to a quite recent period could be embodied in a map of the planet, showing the bright and dark regions of its surface, and in the fact that a white cap would be generally seen to surround each of its poles. When a pole was inclined toward us, and therefore toward the sun, this cap gradually grew smaller, enlarging again when the pole was turned from the sun. In the latter case it would be invisible from the earth, so that the growth would be recognised only by its larger size when it again came into sight. These caps were naturally supposed to be snow and ice which formed around the poles during the Martian winter, and partly or wholly melted away during the summer.

THE CANALS OF MARS

In 1877 commenced Schiaparelli's celebrated observations on the surface of Mars, and his announcement of the so-called canals. The latter consisted of streaks passing from point to point on the planet, and slightly darker than the general surface. Seldom has more misapprehension been caused by a mistranslation than in the present case. Schiaparelli called these streaks *canale*, an Italian word meaning channels. He called them so because it was then supposed that the darker regions of the surface were oceans, and the streams connecting the oceans were therefore supposed to be water, and so were called channels. But the translation "canals" led to a widespread notion that these streaks were the works of inhabitants, as canals on the earth are the works of men.

Up to the present time there is some disagreement between observers and astronomical authorities on the subject of these channels. This arises from the fact that they are not well-defined features on an otherwise uniform surface. Everywhere on the planet are found variations of shade— light and dark patches, so faint and ill-defined that it is generally difficult to assign exact form and outline to them, running into each other by insensible gradations. The extreme difficulty of making them out at all, and the variety of aspects they present under different illuminations and in different states of our atmosphere, has resulted in a great variety of inconsistent delineations of these objects. At one extreme we have the drawings made by the observers at the Lowell Observatory at Flagstaff, Ariz. These show the channels as fine dark lines, so numerous as to form a network covering the greater part of the surface of the planet. In Schiaparelli's map they are rather broad faint bands, not nearly so well defined as in Lowell's drawings. Lowell's channels are much more numerous than those seen by Schiaparelli. We might therefore suppose that all marked by the latter could be identified on Lowell's map. But such is far from being the case; there is only a general resemblance between the features seen at the two stations. One of the most curious features of Lowell's drawings is that the points where the channels cross each other are marked by dark round spots like circular lakes. No such spots as these are shown on Schiaparelli's map.

One of the best marked features of Mars is a large, dark, nearly circular spot, surrounded by white, which is called *Lacus Solis*, or the Lake of the Sun. All observers agree on this. They also agree in a considerable part as to certain faint streaks or channels extending from this lake. But when we go farther we find that they do not agree as to the number of these channels, nor is there an exact agreement as to the surrounding features. It will be interesting to study two drawings of this region made at the Lick Observatory, probably under the best possible conditions, by Campbell and Hussey, respectively.

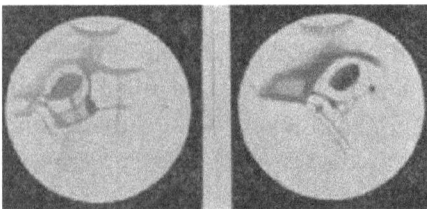

Drawings of Lacus Solis on Mars, by Messrs. Campbell and Hussey

It is not likely that any observatory is more favoured by its atmosphere for observations on this planet than the Lick on Mount Hamilton. Its telescope is the largest and finest in the world that has ever been especially directed to Mars, and Barnard is one of the most cautious observers. It is therefore very noteworthy that on the face of Mars, as presented to Barnard in the Lick telescope, the features do not quite correspond to the channels of Schiaparelli and Lowell. When the air was exceptionally steady he could see a vast number of minute and very faint markings, which were not visible in the smaller telescopes used by the other observers. These were so intricate that it was impossible to represent them on a drawing. They were not confined to the brighter regions of the planet, or the supposed continents, but were found to be more numerous on the so-called seas. They showed no such regularity that they could be considered as channels running from one region to another. The eye could indeed trace darker streaks here and there, and some of these corresponded to the supposed channels, but they were far more irregular than the features on Schiaparelli's and Lowell's maps.

The matter was explained by Cerulli, a careful and industrious Italian observer, in a way which seems very plausible. He found that after he had been studying Mars for two years he was able, by looking at the moon through an opera glass, to see, or fancy he saw, lines and markings upon its surface similar to those of Mars. This phenomenon is not to be regarded as a pure illusion on the one hand, or an exact representation of objects on the other. It grows out of the spontaneous action of the eye in shaping slight and irregular combinations of light and shade, too minute to be separately made out, into regular forms.

PROBABLE NATURE OF THE CHANNELS

The probable facts of the case may be summed up as follows:

1. The surface of Mars is extremely variegated by regions differing in shade, and having no very distinct outlines.

2. There are numerous dark streaks, generally somewhat indefinite in outline, extending through considerable distances across the planet.

3. In many cases the dark portions appear as if chained together to a greater or less extent, and thus give rise to the appearance of long dark channels.

The appearance on which this third phenomenon, which we may regard as identical with that observed by Cerulli, is based, may be well illustrated by looking, with a magnifying glass, at a stippled portrait engraved on steel. Nothing will then be seen but dots, arranged in various lines and curves. But take away the magnifying glass and the eye connects these dots into a well-defined collection of features representing the outlines of the human face. As the eye makes an assemblage of dots into a face, so may it make the minute markings on the planet Mars into the form of long, unbroken channels.

The features which we have hitherto described do not belong to the two polar regions of the planet. Even when the snowcaps have melted away, these regions are seen so obliquely that it would be difficult to trace any well-defined features upon them. The interesting question is whether the caps which cover them are really snow which falls during the Martian winter and melts again when the sun once more shines on the polar regions. To throw light on this question we have to consider some recent results as to the atmosphere of the planet.

THE ATMOSPHERE OF MARS

All recent observers are agreed that, if Mars has any atmosphere at all, it is much rarer than our own, and contains little or no aqueous vapour. This conclusion is reached from observations both with the telescope and the spectroscope. The most careful eye observations of the planet show that the features are rarely, if ever, obscured by anything which can be considered as clouds in the Martian atmosphere. It is true that the features are not always seen with the same distinctness; but the variations in the appearance are no greater than would be due to the changes in the steadiness and purity of our own atmosphere, through which the astronomer necessarily makes his observations. Although, near the edge of the apparent disk of the planet, the features appear to be softened, as if seen through a greater thickness of the atmosphere, this appearance is, at least in part, due to the obliquity of the line of sight, which prevents our getting so good a view of the edge of the disk as of its center. Something of the same sort may be noticed when the moon is viewed with the naked eye or an opera glass. Yet it is quite possible that a certain amount of the softening may be due to a rare atmosphere on Mars.

The most careful spectroscopic examination of the planet was made by Campbell, who compared its spectrum with that of the moon. He could not detect the slightest difference between the two spectra. Now, if Mars had an atmosphere capable of exerting a strong selective absorption on light, we should see lines in the spectrum due to this absorption or, at least, some of the lines would be strengthened. Our general conclusion therefore must be that, while it is quite probable that Mars has an atmosphere, it is one of considerable rarity, and does not bear much aqueous vapour. Now snow can fall only through the condensation of aqueous vapour in the atmosphere. It does not therefore seem likely that much snow can fall on the polar regions of Mars.

Another consideration is that the power of the sun's rays to melt snow is necessarily limited by the amount of heat that they convey. In the polar regions of Mars the rays fall with a great obliquity, and even if all the heat conveyed by them were absorbed, only a few feet of snow could be melted in the course of the summer. But far the larger proportion of this heat must be reflected from the white snow, which is also kept cool by the intense radiation into perfectly cold space. We therefore conclude that the amount of snow that can fall and melt around the polar regions of Mars must be very small, being probably measured by inches at the outside.

As the thinnest fall of snow would suffice to produce a white surface, this does not prove that the caps are not snow. But it seems more likely that the appearance is produced by the simple condensation of aqueous vapour upon the intensely cold surface, producing an appearance similar to that of hoarfrost, which is only frozen dew. This seems to me the most plausible explanation of the polar caps. It has also been suggested that the caps may be due to the condensation of carbonic acid. We can only say of this, that the theory, while not impossible, seems to lack probability.

The reader will excuse me from saying anything in this chapter about the possible inhabitants of Mars. He knows just as much of the subject as I do, and that is nothing at all.

THE SATELLITES OF MARS

No discovery more surprised the whole world than that of two satellites of Mars by Professor Asaph Hall, at the Naval Observatory, in 1877. They had failed of previous detection owing to their extreme minuteness. It was not considered likely that a satellite could be so small as these were found to be, and so no one had taken the trouble to make a careful search with any great telescope. But, when once discovered, they were found to be by no means difficult objects. Of course the ease with which they can be seen depends on the position of Mars both in its orbit and with respect to the earth. They are never visible except when the planet is near its opposition. At each opposition they may be observed for a period of three, four, or even six months, according to circumstances. At an opposition near perihelion they may be seen with a telescope of less than twelve inches diameter; how small a one will show them depends on the skill of the observer, and the pains he takes to cut off the light of the planet from his eye. Generally a telescope ranging from twelve to eighteen inches in diameter is necessary. The difficulty in seeing them arises entirely from the glare of the planet. Could this be eliminated they could doubtless be seen with much smaller instruments. Owing to the glare, the outer one is much easier to see than the inner one, although the inner one is probably the brighter of the two.

Professor Hall assigned the name *Deimos* to the outer and *Phobos* to the inner, these being the attendants of Mars in ancient mythology. Phobos has the remarkable peculiarity that it revolves around the planet in less than nine hours, making its period the shortest of any yet known in the solar system. This is little more than one-third the time of the planet's rotation on its axis. The consequence of this is that, to the inhabitants of the planet, its nearest moon rises in the west and sets in the east.

Deimos performs its revolution in 30 hours 18 minutes. The result of this rapid motion is that some two days must elapse between its rising and setting.

Phobos is only 3700 miles from the surface of the planet. It must therefore be an interesting object to the inhabitants of Mars, if they have telescopes.

In size these bodies are the smallest visible to us in the solar system, with the possible exception of Eros and possibly some others of the fainter asteroids. From Professor Pickering's photometric estimates their diameter was estimated to be not very different from seven miles. Their apparent size as we view them is therefore not very different from that of a small apple hanging over the city of Boston, and seen with a telescope from the city of New York. In this respect they form a singular contrast to nearly or quite all of the other satellites, which are generally a thousand miles or more in diameter. The one exception to this is the fifth satellite of Jupiter. . . . Although this is much less than a thousand miles in diameter, it must considerably exceed the satellites of Mars in size.

The satellites have been most useful to the astronomer in enabling him to learn the exact mass of Mars. How this is done [is explained in Chapter Eight], where the methods of weighing the planets are set forth.

The satellites also offer many curious and difficult problems in gravitation. Their orbits seem to have a slight eccentricity, and the position of the planes in which they revolve changes in consequence of the bulging of the planet at its equator, produced by its rotation. The calculation of these changes and their comparison with observations have opened up a field of research in which Professor Hermann Struve, now of the University of Koenigsberg, Germany, has taken a leading part.

CHAPTER THIRTEEN

The Fairyland of Geometry

Newcomb used the evocative phrase "fairyland of geometry" in a presidential address to the American Mathematical Society in 1897; the speech was reprinted in the journal Science *and in several other journals the following year. He then published a more popular treatment of the subject in* Harper's Magazine *in January 1902. Over the course of the twentieth century, the idea that space might encompass more than the mere three dimensions of everyday experience became a commonplace both of scientific thought and science-fictional speculation. The concept of "compactified" and so unobservable higher dimensions of space rests at the heart of modern-day string theory and its attempt to provide a so-called "theory of everything" to explain the fundamental structure of the physical world. Nonetheless, Newcomb's piece, though more than a century old, remains a clear, concise, and charming introduction to the basics of the subject.*

If the reader were asked in what branch of science the imagination is confined within the strictest limits, he would, I fancy, reply that it must be that of mathematics. The pursuer of this science deals only with problems requiring the most exact statements and the most rigorous reasoning. In all other fields of thought more or less room for play may be allowed to the imagination, but here it is fettered by iron rules, expressed in the most rigid logical form, from which no deviation can be allowed. We are told by philosophers that absolute certainty is unattainable in all ordinary human affairs, the only field in which it is reached being that of geometric demonstration.

And yet geometry itself has its fairyland—a land in which the imagination, while adhering to the forms of the strictest demonstration, roams farther than it ever did in the dreams of Grimm or Andersen. One thing which gives this field its strictly mathematical character is that it was discovered and explored in the search after something to supply an actual want of mathematical science, and was incited by this want rather than by any desire to give play to fancy. Geometricians have always sought to found their science on the most logical basis possible, and thus have carefully and critically inquired into its foundations. The new geometry which has thus arisen is of two closely related yet distinct forms. One of these is called *non-Euclidean*, because Euclid's axiom of parallels, which we shall presently explain, is ignored. In the other form space is assumed to have one or more dimensions in addition to the three to which the space we actually inhabit is confined. As we go beyond the limits set by Euclid in adding a fourth dimension to space, this last branch as well as the other is often designated non-Euclidean. But the more common term is *hypergeometry*, which, though belonging more especially to space of more than three dimensions, is also sometimes applied to any geometric system which transcends our ordinary ideas.

In all geometric reasoning some propositions are necessarily taken for granted. These are called axioms, and are commonly regarded as self-evident. Yet their vital principle is not so much that of being self-evident as being, from the nature of the case, incapable of demonstration. Our edifice must have some support to rest upon, and we take these axioms as its foundation. One example of such a geometric axiom is that only one straight line can be drawn between two fixed points; in other words, two straight lines can never intersect in more than a single point. The axiom with which we are at present concerned is commonly known as the eleventh of Euclid, and may be set forth in the following way: We have given a straight line, AB, and a point, P, with another line, CD, passing through it and capable of being turned around on P. Euclid assumes that this line CD will have one position in which it will be parallel to AB, that is,

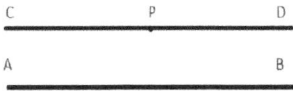

Figure 1

a position such that if the two lines are produced without end, they will never meet. His axiom is that only one such line can be drawn through P. That is to say, if we make the slightest possible change in the direction of the line CD, it will intersect the other line, either in one direction or the other.

The new geometry grew out of the feeling that this proposition ought to be proved rather than taken as an axiom; in fact, that it could in some way be derived from the other axioms. Many demonstrations of it were attempted, but it was always found, on critical examination, that the proposition itself, or its equivalent, had slyly worked itself in as part of the base of the reasoning, so that the very thing to be proved was really taken for granted.

This suggested another course of inquiry. If this axiom of parallels does not follow from the other axioms, then from these latter we may construct a system of geometry in which the axiom of parallels shall not be true. This was done by Lobatchewsky and Bolyai, the one a Russian, the other a Hungarian geometer, about 1830.

To show how a result which looks absurd, and is really inconceivable by us, can be treated as possible in geometry, we must have recourse to analogy. Suppose a world consisting of a boundless flat plane to be inhabited by reasoning beings who can move about at pleasure on the plane, but are not able to turn their heads up or down, or even to see or think of such terms as above them and below them, and things around them can be pushed or pulled about in any direction, but cannot be lifted up. People and things can pass around each other, but cannot step over anything. These dwellers in "flat-land" could construct a plane geometry which would be exactly like ours in being based on the axioms of Euclid. Two parallel straight lines would never meet, though continued indefinitely.

But suppose that the surface on which these beings live, instead of being an infinitely extended plane, is really the surface of an immense globe, like the earth on which we live. It needs no knowledge of geometry, but only an examination of any globular object—an apple, for example—to show that if we draw a line as straight as possible on a sphere, and parallel to it draw a small piece of a second line, and continue this in as straight a line as we can, the two lines will meet when we proceed in either direction one-quarter of the way around the sphere. For our "flat-land" people these lines would both be perfectly straight, because the only curvature would be in the direction downward, which they could never either perceive or discover. The lines would also correspond to the definition of straight lines, because any portion of either contained between two of its points would be the shortest distance between those points. And yet, if these people should extend their measures far enough, they would find any two parallel lines to meet in two points in opposite directions. For all small spaces the axioms of their geometry would apparently hold good, but when they came to spaces as immense as the semi-diameter of the earth, they would find the seemingly absurd result that two parallel lines would, in the course of thousands of miles, come together. Another result yet more astonishing would be that, going ahead far enough in a straight line, they would find that although they had been going forward all the time in what seemed to them the same direction, they would at the end of 25,000 miles find themselves once more at their starting-point.

One form of the modern non-Euclidean geometry assumes that a similar theorem is true for the space in which our universe is contained. Although two straight lines, when continued indefinitely, do not appear to converge even at the immense distances which separate us from the fixed stars, it is possible that there may be a point at which they would eventually meet without either line having deviated from its primitive direction as we understand the case. It would follow that, if we could start out from the earth and fly through space in a perfectly straight line with a velocity perhaps millions of times that of light, we might at length find ourselves approaching the earth from a direction the opposite of that in which we started. Our straight-line circle would be complete.

Another result of the theory is that, if it be true, space, though still unbounded, is not infinite, just as the surface of a sphere, though without any edge or boundary, has only a limited extent of surface. Space would then have only a certain volume—a volume which, though perhaps greater than that of all the atoms in the material universe, would still be capable of being expressed in cubic miles. If we imagine our earth to grow larger and larger in every direction without limit, and with a speed similar to that we have described, so that tomorrow it was large enough to extend to the nearest fixed stars, the day after to yet farther stars, and so on, and we, living upon it, looked out for the result, we should, in time, see the other side of the earth above us, coming down upon us, as it were. The space intervening would grow smaller, at last being filled up. The earth would then be so expanded as to fill all existing space.

This, although to us the most interesting form of the non-Euclidean geometry, is not the only one. The idea which Lobatchewsky worked out was that through a point more than one parallel to a given line could be drawn; that is to say, if through the point P we have already supposed another line were drawn making ever so small an angle with CD, this line also would never meet the line AB. It might approach the latter at first, but would eventually diverge. The two lines AB and CD, starting parallel, would eventually, perhaps at distances greater than that of the fixed stars, gradually diverge from each other. This system does not admit of being shown by analogy so easily as the other, but an idea of it may be had by supposing that the surface of "flat-land," instead of being spherical, is saddle-shaped. Apparently straight parallel lines drawn upon it would then diverge, as supposed by Bolyai. We cannot, however, imagine such a surface extended indefinitely without losing its properties. The analogy is not so clearly marked as in the other case.

To explain hypergeometry proper we must first set forth what a fourth dimension of space means, and show how natural the way is by which it may be approached. We continue our analogy from "flat-land." In this supposed land let us make a cross—two straight lines intersecting at right angles. The inhabitants of this land understand the cross perfectly, and conceive of it just as we do. But let us ask them to draw a third line, intersecting in the same point, and perpendicular to both the other lines. They would at once pronounce this absurd and impossible. It is equally absurd and impossible to us if we require the third line to be drawn on the paper. But we should reply, "If you allow us to leave the paper or flat surface, then we can solve the problem by simply drawing the third line through the paper perpendicular to its surface" (Figure 2).

Now, to pursue the analogy, suppose that, after we have drawn three mutually perpendicular lines, some being from another sphere proposes to us the drawing of a fourth line through the same point, perpendicular to all three of the lines already there. We should answer him in the same way that the inhabitants of "flat-land" answered us: "The problem is impossible. You cannot draw any such line in space as we understand it." If our visitor conceived of the fourth dimension, he would reply to us as we replied to the "flat-land" people: "The problem is absurd and impossible if you confine your line to space as you understand it. But for me there is a fourth dimension in space. Draw your line through that dimension, and the problem will be solved. This is perfectly simple to me; it is impossible to you solely because your conceptions do not admit of more than three dimensions."

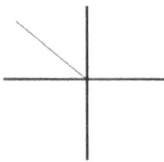

Figure 2

Supposing the inhabitants of "flat-land" to be intellectual beings as we are, it would be interesting to them to be told what dwellers of space in three dimensions could do. Let us pursue the analogy by showing what dwellers in four dimensions might do. Place a dweller of "flat-land" inside a circle drawn on his plane, and ask him to step outside of it without breaking through it. He would go all around, and, finding every inch of it closed, he would say it was impossible from the very nature of the conditions. "But," we would reply, "that is because of your limited conceptions. We can step over it."

"Step over it!" he would exclaim. "I do not know what that means. I can pass around anything if there is a way open, but I cannot imagine what you mean by stepping over it."

But we should simply step over the line and reappear on the other side. So, if we confine a being able to move in a fourth dimension in the walls of a dungeon of which the sides, the floor, and the ceiling were all impenetrable, he would step outside of it without touching any part of the building, just as easily as we could step over a circle drawn on the plane without

Figure 3

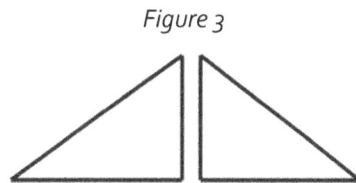

touching it. He would simply disappear from our view like a spirit, and perhaps reappear the next moment outside the prison. To do this he would only have to make a little excursion in the fourth dimension.

Another curious application of the principle is more purely geometrical. We have here two triangles, of which the sides and angles of the one are all equal to corresponding sides and angles of the other (Figure 3). Euclid takes it for granted that the one triangle can be laid upon the other so that the two shall fit together. But this cannot be done unless we lift one up and turn it over. In the geometry of "flat-land" such a thing as lifting up is inconceivable; the two triangles could never be fitted together.

Now let us suppose two pyramids similarly related (Figure 4). All the faces and angles of the one correspond to the faces and angles of the other. Yet, lift them about as we please, we could never fit them together. If we fit

Figure 4

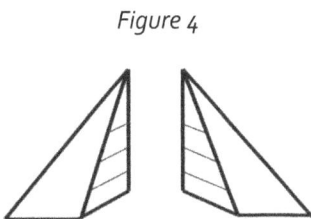

the bases together the two will lie on opposite sides, one being below the other. But the dweller in four dimensions of space will fit them together without any trouble. By the mere turning over of one he will convert it into the other without any change whatever in the relative position of its parts. What he could do with the pyramids he could also do

103

with one of us if we allowed him to take hold of us and turn a somersault with us in the fourth dimension. We should then come back into our natural space, but changed as if we were seen in a mirror. Everything on us would be changed from right to left, even the seams in our clothes, and every hair on our head. All this would be done without, during any of the motion, any change having occurred in the positions of the parts of the body.

It is very curious that, in these transcendental speculations, the most rigorous mathematical methods correspond to the most mystical ideas of the Swedenborgian and other forms of religion. Right around us, but in a direction which we cannot conceive any more than the inhabitants of "flat-land" can conceive up and down, there may exist not merely another universe, but any number of universes. All that physical science can say against the supposition is that, even if a fourth dimension exists, there is some law of all the matter with which we are acquainted which prevents any of it from entering that dimension, so that, in our natural condition, it must forever remain unknown to us.

Another possibility in space of four dimensions would be that of turning a hollow sphere, an India rubber ball, for example, inside out by simple bending without tearing it. To show the motion in our space to which this is analogous, let us take a thin, round sheet of India rubber, and cut out all the central part, leaving only a narrow ring round the border. Suppose the outer edge of this ring fastened down on a table, while we take hold of the inner edge and stretch it upward and outward over the outer edge until we flatten the whole ring on the table, upside down, with the inner edge now the outer one. This motion would be as inconceivable in "flat-land" as turning the ball inside out is to us.

CHAPTER FOURTEEN

The Relation of Scientific Method
to Social Progress

Newcomb was not afraid to speak and write on subjects beyond mathematics and astronomy, including politics, economics, and social issues in general. This was no doubt one of the factors that made him perhaps the best-known American scientist of the late nineteenth and early twentieth century. Like many scientists before and since—Carl Sagan and Isaac Asimov come to mind as proponents of the same idea— Newcomb believed that the scientific method, if properly applied to other realms of social life, could play a major role in improving the human condition. He made his case in his presidential address to the Philosophical Society of Washington; the piece was later reprinted in Side-Lights on Astronomy.

Among those subjects which are not always correctly apprehended, even by educated men, we may place that of the true significance of scientific method and the relations of such method to practical affairs. This is especially apt to be the case in a country like our own, where the points of contact between the scientific world on the one hand, and the industrial and political world on the other, are fewer than in other civilized countries. The form which this misapprehension usually takes is that of a failure to appreciate the character of scientific method, and especially its analogy to the methods of practical life. In the judgment of the ordinary intelligent man there is a wide distinction between theoretical and practical science. The latter he considers as that science directly applicable to the building of railroads, the construction of engines, the invention of new machinery, the construction of maps, and other useful objects. The former he considers analogous to those philosophic speculations in which men have indulged in all ages without leading to any result which he considers practical. That our knowledge of nature is increased by its prosecution is a fact of which he is quite conscious, but he considers it as terminating with a mere increase of knowledge, and not as having in its method anything which a person devoted to material interests can be expected to appreciate.

This view is strengthened by the spirit with which he sees scientific investigation prosecuted. It is well understood on all sides that when such investigations are pursued in a spirit really recognized as scientific, no merely utilitarian object is had in view. Indeed, it is easy to see how the very fact of pursuing such an object would detract from that thoroughness of examination which is the first condition of a real advance. True science demands in its every research a completeness far beyond what is apparently necessary for its practical applications. The precision with which the astronomer seeks to measure the heavens and the chemist to determine the relations of the ultimate molecules of matter has no limit, except that set by the imperfections of the instruments of research. There is no such division recognized as that of useful and useless knowledge. The ultimate aim is nothing less than that of bringing all the phenomena of nature under laws as exact as those which govern the planetary motions.

Now the pursuit of any high object in this spirit commands from men of wide views that respect which is felt towards all exertion having in view more elevated objects than the pursuit of gain. Accordingly, it is very natural to classify scientists and philosophers with the men who in all ages have sought after learning instead of utility. But there is another aspect of the question which will show the relations of scientific advance to the practical affairs of life in a different light. I make bold to say that the greatest want of the day, from a purely practical point of view, is the more general introduction of the scientific method and the scientific spirit into the discussion of those political and social problems which we encounter on our road to a higher plane of public well being. Far from using methods too refined for practical purposes, what most distinguishes scientific from other thought is the introduction of the methods of practical life into the discussion of abstract general problems. A single instance will illustrate the lesson I wish to enforce.

The question of the tariff is, from a practical point of view, one of the most important with which our legislators will have to deal during the next few years. The widest diversity of opinion exists as to the best policy to be pursued in collecting a revenue from imports. Opposing interests contend against one another without any common basis of fact or principle on which a conclusion can be reached. The opinions of intelligent men differ almost as widely as those of the men who are immediately interested. But all will admit that public action in this direction should be dictated by one guiding principle—that the greatest good of the community is to be sought after. That policy is the best which will most promote this good. Nor is there any serious difference of opinion as to the nature of the good to be had in view; it is in a word the increase of the national wealth and prosperity. The question on which opinions fundamentally differ is that of the effects of a higher or lower rate of duty upon the interests of the public. If it were possible to foresee, with an approach to certainty, what effect a given tariff would have upon the producers and consumers of an article taxed, and, indirectly, upon each member of the community in any way interested in the article, we should then have an exact datum which we do not now possess for reaching a conclusion. If some superhuman authority, speaking with the voice of infallibility, could give us this information, it is evident that a great national want would be supplied. No question in practical life is more important than this: How can this desirable knowledge of the economic effects of a tariff be obtained?

The answer to this question is clear and simple. The subject must be studied in the same spirit, and, to a certain extent, by the same methods which have been so successful in advancing our knowledge of nature. Everyone knows that, within the last two centuries, a method of studying the course of nature has been introduced which has been so successful in enabling us to trace the sequence of cause and effect as almost to revolutionize society. The very fact that scientific method has been so successful here leads to the belief that it might be equally successful in other departments of inquiry.

The same remarks will apply to the questions connected with banking and currency; the standard of value; and, indeed, all subjects which have a financial bearing. On every such question we see wide differences of opinion without any common basis to rest upon.

It may be said, in reply, that in these cases there are really no grounds for forming an opinion, and that the contests which arise over them are merely those between conflicting interests. But this claim is not at all consonant with the form which we see the discussion assume. Nearly everyone has a decided opinion on these several subjects; whereas, if there were no data for forming an opinion, it would be unreasonable to maintain any whatever. Indeed, it is evident that there must be truth somewhere, and the only question that can be open is that of the mode of discovering it. No man imbued with a scientific spirit can claim that such truth is beyond the power of the human intellect. He may doubt his own ability to grasp it, but cannot doubt that by pursuing the proper method and adopting the best means the problem can be solved. It is, in fact, difficult to show why some exact results could not be as certainly reached in economic questions as in those of physical science. It is true that if we pursue the inquiry far enough we shall find more complex conditions to encounter, because the future course of demand and supply enters as an uncertain element. But a remarkable fact to be considered is that the difference of opinion to which we allude does not depend upon different estimates of the future, but upon different views of the most elementary and general principles of the subject. It is as if men were not agreed whether air were elastic or whether the earth turns on its axis. Why is it that while in all subjects of physical science we find a general agreement through a wide range of subjects, and doubt commences only where certainty is not attained, yet when we turn to economic subjects we do not find the beginning of an agreement?

No two answers can be given. It is because the two classes of subjects are investigated by different instruments and in a different spirit. The physicist has an exact nomenclature; uses methods of research well adapted to the objects he has in view; pursues his investigations without being attacked by those who wish for different results; and, above all, pursues them only for the purpose of discovering the truth. In economic questions the case is entirely different. Only in rare cases are they studied without at least the suspicion that the student has a preconceived theory to support. If results are attained which oppose any powerful interest, this interest can hire a competing investigator to bring out a different result. So far as the public can see, one man's result is as good as another's, and thus the object is as far off as ever. We may be sure that until there is an intelligent and rational public, able to distinguish between the speculations of the charlatan and the researches of the investigator, the present state of things will continue. What we want is so wide a diffusion of scientific ideas that there shall be a class of men engaged in studying economic problems for their own sake, and an intelligent public able to judge what they are doing. There must be an improvement in the objects at which they aim in education, and it is now worthwhile to inquire what that improvement is.

It is not mere instruction in any branch of technical science that is wanted. No knowledge of chemistry, physics, or biology, however extensive, can give the learner much aid in forming a correct opinion of such a question as that of the currency. If we should claim that political economy ought to be more extensively studied, we would be met by the question, which of several conflicting systems shall we teach? What is wanted is not to teach this system or that, but to give such a training that the student shall be able to decide for himself which system is right.

It seems to me that the true educational want is ignored both by those who advocate a classical and those who advocate a scientific education. What is really wanted is to train the intellectual powers, and the question ought to be, what is the best method of doing this? Perhaps it might be found that both of the conflicting methods could be improved upon. The really distinctive features, which we should desire to see introduced, are two in number: the one the scientific spirit; the other the scientific discipline. Although many details may be classified under each of these heads, yet there is one of pre-eminent importance on which we should insist.

The one feature of the scientific spirit which outweighs all others in importance is the love of knowledge for its own sake. If by our system of education we can inculcate this sentiment we shall do what is, from a public point of view, worth more than any amount of technical knowledge, because we shall lay the foundation of all knowledge. So long as men study only what they think is going to be useful their knowledge will be partial and insufficient. I think it is to the constant inculcation of this fact by experience, rather than to any reasoning, that is due the continued appreciation of a liberal education. Every businessman knows that a business-college training is of very little account in enabling one to fight the battle of life, and that college-bred men have a great advantage even in fields where mere education is a secondary matter. We are accustomed to seeing ridicule thrown upon the questions sometimes asked of candidates for the civil service because the questions refer to subjects of which a knowledge is not essential. The reply to all criticisms of this kind is that there is no one quality which more certainly assures a man's usefulness to society than the propensity to acquire useless knowledge. Most of our citizens take a wide interest in public affairs, else our form of government would be a failure. But it is desirable that their study of public measures should be more critical and take a wider range. It is especially desirable that the conclusions to which they are led should be unaffected by partisan sympathies. The more strongly the love of mere truth is inculcated in their nature the better this end will be attained.

The scientific discipline to which I ask mainly to call your attention consists in training the scholar to the scientific use of language. Although whole volumes may be written on the logic of science there is one general feature of its method which is of fundamental significance. It is that every term which it uses and every proposition which it enunciates has a precise meaning which can be made evident by proper definitions. This general principle of scientific language is much more easily inculcated by example than subject to exact description; but I shall ask leave to add one to several attempts I have made to define it. If I should say that when a statement is made in the language of science the speaker knows what he means, and the hearer either knows it or can be made to know it by proper definitions, and that this community of understanding is frequently not reached in other departments of thought, I might be understood as casting a slur on whole departments of inquiry. Without intending any such slur, I may still say that language and statements are worthy of the name scientific as they approach this standard; and, moreover, that a great deal is said and written which does not fulfil the requirement. The fact that words lose their meaning when removed from the connections in which that meaning has been acquired and put to higher uses, is one which, I think, is rarely recognized. There is nothing in the history of philosophical inquiry more curious than the frequency of interminable disputes on subjects where no agreement can be reached because the opposing parties do not use words in the same sense. That the history of science is not free from this reproach is shown by the fact of the long dispute whether the force of a moving body was proportional to the simple velocity or to its square. Neither of the parties to the dispute thought it worthwhile to define what they meant by the word "force," and it was at length found that if a definition was agreed upon the seeming difference of opinion would vanish. Perhaps the most striking feature of the case, and one peculiar to a scientific dispute, was that the opposing parties did not differ in their solution of a single mechanical problem. I say this is curious, because the very fact of their agreeing upon every concrete question which could have been presented ought to have made it clear that some fallacy was lacking in the discussion as to the measure of force. The good effect of a scientific spirit is shown by the fact that this discussion is almost unique in the history of science during the past two centuries, and that scientific men themselves were able to see the fallacy involved, and thus to bring the matter to a conclusion.

If we now turn to the discussion of philosophers, we shall find at least one yet more striking example of the same kind. The question of the freedom of the human will has, I believe, raged for centuries. It cannot yet be said that any conclusion has been reached. Indeed, I have heard it admitted by men of high intellectual attainments that the question was insoluble. Now a curious feature of this dispute is that none of the combatants, at least on the affirmative side, have made any serious attempt to define what should be meant by the phrase freedom of the will, except by using such terms as require definition equally with the word freedom itself. It can, I conceive, be made quite clear that the assertion, "The will is free," is one without meaning, until we analyze more fully the different meanings to be attached to the word free. Now this word has a perfectly well-defined signification in everyday life. We say that anything is free when it is not subject to external constraint. We also know exactly what we mean when we say that a man is free to do a certain act. We mean that if he chooses to do it there is no external constraint acting to prevent him. In all cases a relation of two things is implied in the word, some active agent or power, and the presence or absence of another constraining agent. Now, when we inquire whether the will itself is free, irrespective of external constraints, the word free no longer has a meaning, because one of the elements implied in it is ignored.

To inquire whether the will itself is free is like inquiring whether fire itself is consumed by the burning, or whether clothing is itself clad. It is not, therefore, at all surprising that both parties have been able to dispute without end, but it is a most astonishing phenomenon of the human intellect that the dispute should go on generation after generation without the parties finding out whether there was really any difference of opinion between them on the subject. I venture to say that if there is any such difference, neither party has ever analyzed the meaning of the words used sufficiently far to show it. The daily experience of every man, from his cradle to his grave, shows that human acts are as much the subject of external causal influences as are the phenomena of nature. To dispute this would be little short of the ludicrous. All that the opponents of freedom, as a class, have ever claimed is the assertion of a causal connection between the acts of the will and influences independent of the will. True, propositions of this sort can be expressed in a variety of ways connoting an endless number of more or less objectionable ideas, but this is the substance of the matter.

To suppose that the advocates on the other side meant to take issue on this proposition would be to assume that they did not know what they were saying. The conclusion forced upon us is that though men spend their whole lives in the study of the most elevated department of human thought it does not guard them against the danger of using words without meaning. It would be a mark of ignorance, rather than of penetration, to hastily denounce propositions on subjects we are not well acquainted with because we do not understand their meaning. I do not mean to intimate that philosophy itself is subject to this reproach. When we see a philosophical proposition couched in terms we do not understand, the most modest and charitable view is to assume that this arises from our lack of knowledge. Nothing is easier than for the ignorant to ridicule the propositions of the learned. And yet, with every reserve, I cannot but feel that the disputes to which I have alluded prove the necessity of bringing scientific precision of language into the whole domain of thought. If the discussion had been confined to a few, and other philosophers had analyzed the subject, and showed the fictitious character of the discussion, or had pointed out where opinions really might differ, there would be nothing derogatory to philosophers. But the most suggestive circumstance is that although a large proportion of the philosophic writers in recent times have devoted more or less attention to the subject, few, or none, have made even this modest contribution. I speak with some little confidence on this subject, because several years ago I wrote to one of the most acute thinkers of the country, asking if he could find in philosophic literature any terms or definitions expressive of the three different senses in which not only the word freedom, but nearly all words implying freedom were used. His search was in vain.

Nothing of this sort occurs in the practical affairs of life. All terms used in business, however general or abstract, have that well-defined meaning which is the first requisite of the scientific language. Now one important lesson which I wish to inculcate is that the language of science in this respect corresponds to that of business; in that each and every term that is employed has a meaning as well defined as the subject of discussion can admit of. It will be an instructive exercise to inquire what this peculiarity of scientific and business language is. It can be shown that a certain requirement should be fulfilled by all language intended for the discovery of truth, which is fulfilled only by the two classes of language which I have described. It is one of the most common errors of discourse to assume that any common expression which we may use always conveys an idea, no matter what the subject of discourse. The true state of the case can, perhaps, best be seen by beginning at the foundation of things and examining under what conditions language can really convey ideas.

Suppose thrown among us a person of well-developed intellect, but un-acquainted with a single language or word that we use. It is absolutely useless to talk to him, because nothing that we say conveys any meaning to his mind. We can supply him no dictionary, because by hypothesis he knows no language to which we have access. How shall we proceed to communicate our ideas to him? Clearly there is but one possible way—namely, through his senses. Outside of this means of bringing him in contact with us we can have no communication with him. We, therefore, begin by showing him sensible objects, and letting him understand that certain words which we use correspond to those objects. After he has thus acquired a small vocabulary, we make him understand that other terms refer to relations between objects which he can perceive by his senses. Next he learns, by induction, that there are terms which apply not to special objects, but to whole classes of objects. Continuing the same process, he learns that there are certain attributes of objects made known by the manner in which they affect his senses, to which abstract terms are applied. Having learned all this, we can teach him new words by combining words without exhibiting objects already known. Using these words we can proceed yet further, building up, as it were, a complete language. But there is one limit at every step. Every term which we make known to him must depend ultimately upon terms the meaning of which he has learned from their connection with special objects of sense.

To communicate to him a knowledge of words expressive of mental states it is necessary to assume that his own mind is subject to these states as well as our own, and that we can in some way indicate them by our acts. That the former hypothesis is sufficiently well established can be made evident so long as a consistency of different words and ideas is maintained. If no such consistency of meaning on his part were evident, it might indicate that the operations of his mind were so different from ours that no such communication of ideas was possible. Uncertainty in this respect must arise as soon as we go beyond those mental states which communicate themselves to the senses of others.

We now see that in order to communicate to our foreigner a knowledge of language, we must follow rules similar to those necessary for the stability of a building. The foundation of the building must be well laid upon objects knowable by his five senses. Of course the mind, as well as the external object, may be a factor in determining the ideas which the words are intended to express; but this does not in any manner invalidate the conditions which we impose. Whatever theory we may adopt of the relative part played by the knowing subject, and the external object in the acquirement of knowledge, it remains none the less true that no knowledge of the meaning of a word can be acquired except through the senses, and that the meaning is, therefore, limited by the senses. If we transgress the rule of founding each meaning upon meanings below it, and having the whole ultimately resting upon a sensuous foundation, we at once branch off into sound without sense. We may teach him the use of an extended vocabulary, to the terms of which he may apply ideas of his own, more or less vague, but there will be no way of deciding that he attaches the same meaning to these terms that we do.

What we have shown true of an intelligent foreigner is necessarily true of the growing child. We come into the world without a knowledge of the meaning of words, and can acquire such knowledge only by a process which we have found applicable to the intelligent foreigner. But to confine ourselves within these limits in the use of language requires a course of severe mental discipline. The transgression of the rule will naturally seem to the undisciplined mind a mark of intellectual vigor rather than the reverse. In our system of education every temptation is held out to the learner to transgress the rule by the fluent use of language to which it is doubtful if he himself attaches clear notions, and which he can never be certain suggests to his hearer the ideas which he desires to convey. Indeed, we not infrequently see, even among practical educators, expressions of positive antipathy to scientific precision of language so obviously opposed to good sense that they can be attributed only to a failure to comprehend the meaning of the language which they criticize.

Perhaps the most injurious effect in this direction arises from the natural tendency of the mind, when not subject to a scientific discipline, to think of words expressing sensible objects and their relations as connoting certain supersensuous attributes. This is frequently seen in the repugnance of the metaphysical mind to receive a scientific statement about a matter of fact simply as a matter of fact. This repugnance does not generally arise in respect to the everyday matters of life. When we say that the earth is round we state a truth which everyone is willing to receive as final. If without denying that the earth was round, one should criticise the statement on the ground that it was not necessarily round but might be of some other form, we should simply smile at this use of language. But when we take a more general statement and assert that the laws of nature are inexorable, and that all phenomena, so far as we can show, occur in obedience to their requirements, we are met with a sort of criticism with which all of us are familiar, but which I am unable adequately to describe. No one denies that as a matter of fact, and as far as his experience extends, these laws do appear to be inexorable. I have never heard of any one professing, during the present generation, to describe a natural phenomenon, with the avowed belief that it was not a product of natural law; yet we constantly hear the scientific view criticised on the ground that events *may* occur without being subject to natural law. The word "may," in this connection, is one to which we can attach no meaning expressive of a sensuous relation.

The analogous conflict between the scientific use of language and the use made by some philosophers is found in connection with the idea of causation. Fundamentally the word cause is used in scientific language in the same sense as in the language of common life. When we discuss with our neighbors the cause of a fit of illness, of a fire, or of cold weather, not the slightest ambiguity attaches to the use of the word, because whatever meaning may be given to it is founded only on an accurate analysis of the ideas involved in it from daily use. No philosopher objects to the common meaning of the word, yet we frequently find men of eminence in the intellectual world who will not tolerate the scientific man in using the word in this way. In every explanation which he can give to its use they detect ambiguity. They insist that in any proper use of the term the idea of power must be connoted. But what meaning is here attached to the word power, and how shall we first reduce it to a sensible form, and then apply its meaning to the operations of nature? Whether this can be done, I do not inquire. All I maintain is that if we wish to do it, we must pass without the domain of scientific statement.

Perhaps the greatest advantage in the use of symbolic and other mathematical language in scientific investigation is that it cannot possibly be made to connote anything except what the speaker means. It adheres to the subject matter of discourse with a tenacity which no criticism can overcome. In consequence, whenever a science is reduced to a mathematical form its conclusions are no longer the subject of philosophical attack. To secure the same desirable quality in all other scientific language it is necessary to give it, so far as possible, the same simplicity of signification which attaches to mathematical symbols. This is not easy, because we are obliged to use words of ordinary language, and it is impossible to divest them of whatever they may connote to ordinary hearers.

I have thus sought to make it clear that the language of science corresponds to that of ordinary life, and especially of business life, in confining its meaning to phenomena. An analogous statement may be made of the method and objects of scientific investigation. I think Professor Clifford was very happy in defining science as organized common-sense. The foundation of its widest general creations is laid, not in any artificial theories, but in the natural beliefs and tendencies of the human mind. Its position against those who deny these generalizations is quite analogous to that taken by the Scottish school of philosophy against the skepticism of Hume.

It may be asked, if the methods and language of science correspond to those of practical life, why is not the everyday discipline of that life as good as the discipline of science? The answer is, that the power of transferring the modes of thought of common life to subjects of a higher order of generality is a rare faculty which can be acquired only by scientific discipline. What we want is that in public affairs men shall reason about questions of finance, trade, national wealth, legislation, and administration, with the same consciousness of the practical side that they reason about their own interests. When this habit is once acquired and appreciated, the scientific method will naturally be applied to the study of questions of social policy. When a scientific interest is taken in such questions, their boundaries will be extended beyond the utilities immediately involved, and one important condition of unceasing progress will be complied with.

CHAPTER FIFTEEN

Is the Airship Coming?

Ironically, given his long and productive scientific career, perhaps the one thing Newcomb is best known for today, more than a century after his death, is a prediction—a failed prediction. Newcomb was by no means the only scientist to pooh-pooh the possibility of heavier-than-air flying machines (as opposed to balloons and dirigibles), but he was probably the most eminent. The reasons for Newcomb's lack of foresight are interesting, and I discuss them further in the epilogue to this book. But if Newcomb was right in regard to the Martian canals, it cannot be denied that he was wrong when it came to airplanes. Two years and two months after "Is the Airship Coming?" appeared in McClure's Magazine, *the Wright brothers made their first flight at Kitty Hawk.*

I s the airship coming? That depends, first of all, on whether we are able to make the requisite scientific discoveries. To do this we must penetrate a field of thought which Nature has hitherto held sacred from the tread of the most adventurous scientific explorer. What the human mind has been able to grasp belongs to a middle class of things between the infinitely great and the infinitely small. The universe made known to us by the telescope requires so many eons to go through a single stage of its growth that its origin and destiny are alike incomprehensible to a being who can observe it but for a few fleeting years. Its vastness defies comprehension and eludes investigation. The microscope has made known to us how active and busy a world may be bounded by the surface of a single drop of water, in whose crystal purity the unaided eye cannot distinguish a speck. But the power of the microscope has a limit set by the very nature of light itself. Far below that limit, within a single cell so minute as almost to elude vision, even with the most powerful microscope, are the myriad molecules of matter. We have evidence that a single one of these, so minute that its individual existence could never be made known to us by any process whatever, is a mechanism whose complexity evades description—a seat of activity going through cycles of change millions of times in the millionth of a second.

Between these extremes lie two connecting links—invisible bonds, making known their existence to our universal experience, and yet evading investigation, as do the infinitely great and the infinitely small. They are the luminiferous ether and the force of gravitation. The former, invisible and imperceptible as the optic nerve is imperceptible to its own sight, fills all space. And yet, its minutest parts are susceptible to the vibrations of light which number hundreds of millions of millions in a single second and are propagated with such speed as to fly to the moon in two beats of the clock. Gravitation binds every separate molecule of matter on the earth to every molecule on the planets and every molecule in the most distant star. Yet up to the present time the profoundest philosopher knows no more about its why and wherefore, if why and wherefore it has, than the child that knows it will fall to the ground if its foot slips.

It goes without saying that the science of today is not satisfied to accept any of these limitations longer than it is forced to do so. It is battering at every gate which Nature has closed against the entrance of its forces. Well knowing that the eye of man is never to see a molecule of matter, it is nevertheless investigating the phenomena associated with it, determined, if possible, to penetrate the mystery of its constitution. It is seeking to discover the cause of gravitation, the force which, coextensive with ether itself, may be in close association with it. From time to time philosophers fancy the road open to success, yet nothing that can be practically called success has yet been reached or even approached. When it is reached, when we are able to state exactly why matter gravitates, then will arise the question how this hitherto unchangeable force may be controlled and regulated. With this question answered the problem of the interaction between ether and matter may be solved. That interaction goes on between ether and molecules is shown by the radiation of heat by all bodies. When the molecules are combined into a mass, this interaction ceases, so that the lightest objects fly through the ether without resistance. Why is this? Why does ether act on the molecule and not the mass? When we can produce the latter, and when the mutual action can be controlled, then may gravitation be overcome and then may men build, not merely airships, but ships which shall fly above the air, and transport their passengers from continent to continent with the speed of the celestial motions.

The first question suggested to the reader by these considerations is whether any such result is possible; whether it is within the power of man to discover the nature of luminiferous ether and the cause of gravitation. To this the profoundest philosopher can only answer, "I do not know." Quite possibly the gates at which he is beating are, in the very nature of things, incapable of being opened. It may be that the mind of man is incapable of grasping the secrets within them. The question has even occurred to me whether, if a being of such supernatural power as to understand the operations going on in a molecule of matter or in a current of electricity as we understand the operations of a steam engine should essay to explain them to us, he would meet with any more success than we should in explaining to a fish the engines of a ship which so rudely invades its domain. As was remarked by William K. Clifford, perhaps the clearest spirit that has ever studied such problems, it is possible that the laws of geometry for spaces infinitely small may be so different from those of larger spaces that we must necessarily be unable to conceive them.

Let us now take up the question from a more immediately practical point of view. Can we decide whether the airship is or is not possible simply as a triumph of invention, unaided by any such revolutionary discovery as that we have suggested? If I should answer no, I should be at once charged with setting limits to the powers of invention, and have held before my eyes, as a warning example, the names of more than one philosopher who has declared things impossible which were afterward brought to pass. Instead of answering yes or no, I shall ask the reader to bear with me while I point out some general features of the progress of science and invention.

Invention and discovery have, notwithstanding their seemingly wide extent, gone on in rather narrower lines than is commonly supposed. If, a hundred years ago, the most sagacious of mortals had been told that before the nineteenth century closed the face of the earth would be changed, time and space almost annihilated, and communication between continents made more rapid and easy than it was between cities in his time; and if he had been asked to exercise his wildest imagination in depicting what might come—the airship and the flying machine would probably have had a prominent place in his scheme, but neither the steamship, the railway, the telegraph, nor the telephone would have been there. Probably not a single new agency which he could have imagined would be one that has come to pass.

For thousands of years mathematicians vainly grappled with three problems : to describe a square which should be equal in area to a given circle, to trisect an angle, to construct a cube having double the solid contents of a given cube. Vastly more complex problems had been solved, why should these evade their powers? When such advances in mathematical thought and methods were reached that the possibility of a solution could be inquired into, the answer was a negative one. That none of these problems could be solved was demonstrated by a process as rigorous, though not so accessible to the ordinary mind, as any proposition in Euclid. But this did not mean that mathematics had ceased to advance. The very demonstration of the impossibility was a triumph second only to what the solution itself would have been.

Great indeed have been the advances in our knowledge of the heavens. And yet, any inquirer can ask the astronomer questions to which he can only answer by avowing his ignorance. He has discovered revolving around the stars worlds which must remain forever invisible, even to the telescopic eye, and can tell what gases are bursting out from some blazing object in the most distant regions of the universe, as readily as a chemist can tell the composition of the most ordinary substance. And yet he remains dumb when asked about the surface of Jupiter, or called upon to tell the inquirer whether Mars is inhabited.

It is so with invention. The distinction between the possible and the impossible is not clear. A useful result may look entirely feasible on such consideration as we can give it, when, if we inquire into the case, we should see an absurdity in expecting it. Not many years ago the public was so much interested in the question of making it rain that Congress provided means to send a party all the way to Texas to see if rain could be brought down by bombarding the skies with dynamite bombs. The incredulous scientists who declared the attempt absurd were held up to ridicule by ardent spirits, while men of more caution held that the experiment was worthy of a trial, even if the chances of success were small.

Now, compare this problem with another, quite similar in principle. Every man who favored an attempt to bring down rain would have ridiculed a proposal to make it high tide in New York Harbor by blowing up the water with dynamite whenever a great ship had to go to sea. Why expect the one and not the other? The answer would be that the one attempt was simply ridiculous while the other was not. No hydrographer would ever be expected to change the course of the Gulf Stream, or to vary its temperature. But it only needs a wide grasp of the subject to see that the problem of bringing to New York from some vaporous region an air current which shall deposit its moisture on an arid field is of the same kind as the problem of changing the Gulf Stream, or bringing a tidal wave into New York Harbor. Nor is it evident that to expect the air to condense its moisture without the necessary conditions being produced would be like expecting an engine to run without fuel.

No builder of air castles for the amusement and benefit of humanity could have failed to include a flying machine among the productions of his imagination. The desire to fly like a bird is inborn in our race, and we can no more be expected to abandon the idea than the ancient mathematician could have been expected to give up the problem of squaring the circle. The lesson which we draw from this general review of progress is that we cannot conclude that because the genius of the nineteenth century has opened up such wonders as it has, therefore the twentieth is to give us the airship. But even granting the abstract possibility of the flying machine or the airship, we are still met with the question of its usefulness as a means of international communication. It would, of course, be very pleasant for a Bostonian who wished to visit New York to take out his wings from the corner of his vestibule, mount them, and fly to the Metropolis. But it is hardly conceivable that he would get there any more quickly or cheaply than he now does by rail.

Another feature incidental to any aerial vehicle is very generally overlooked. In the absence of any such revolutionary discovery as I have pictured in the first part of this article—in the absence of the power to control gravitation—a flying machine could remain in the air only by the action of its machinery, and would fall to the ground like a wounded bird the moment any accident stopped it. With all the improvements that the genius of man has made in the steamship, the greatest and best ever constructed is liable now and then to meet with accident. When this happens she simply floats on the water until the damage is repaired, or help reaches her. Unless we are to suppose for the flying machine, in addition to everything else, an immunity from accident which no human experience leads us to believe possible, it would be liable to derangements of machinery, any one of which would be necessarily fatal. If an engine were necessary not only to propel a ship, but also to make her float—if, on the occasion of any accident she immediately went to the bottom with all on board—there would not, at the present day, be any such thing as steam navigation.

Let us look at the problem and see what room there is for the airship among the inventions of the future. If we are to have an aerial machine of any kind, it must be one of two principles. Either we must control the law of universal gravitation, as I have already suggested, or the machine must be supported by the air.

Only two systems of air-support seem possible, or have ever been suggested. The vehicle must either float in the air, like a balloon, or it must be supported by the action of the air on moving wings, like a bird when it flies. The conditions of both of these methods can be made the subject of exact investigation. A floating vehicle to carry a certain weight must have a bulk corresponding to the volume of air which shall have this weight. With this bulk it must experience a certain resistance to its passage through the air, which resistance increases at least as the square of the velocity. To overcome this resistance requires a corresponding power to be exerted by an engine of some kind. The engine has weight. The best combination of all these conditions is a problem of applied science, of which the solution depends mainly on the strength and weight of materials. Solve it as we will, our floating ship must have a thousand times the bulk of a railroad train carrying an equal weight, and experience a hundred times the resistance that the train does. It therefore seems quite evident that while the problem of a dirigible balloon may be within the power of inventive genius, we cannot hope that it will become a vehicle for carrying passengers and freight under ordinary conditions.

Now let us turn to the other alternative, that of the flying machine. If we can make a model of a bird with its wings, and set the wings in motion like those of a bird with no greater weight, the model will fly like a bird. To do this is, in a certain sense, a problem of nothing but applied mechanics. Yet it has its well-defined limitations. By experiments on the resistance of the air we can compute how large a wing or aeroplane, moving with a certain speed, will be required to support a given weight. We can also determine, or, at least, form some idea of, the power of the engine that will move the apparatus. There must be connecting machinery, by which the engine shall in some way act on the plane. Engine, machinery, and plane must all have a weight proportioned to, or at least increasing with, their size and efficiency. It is then a problem of strength of materials to form a combination in which the ratio of efficiency to weight will be enough to make the machine fly.

In studying the best combination, we meet two difficulties, one of which can be stated in a very simple mathematical form. Let us make two flying machines exactly alike, only make one on double the scale of the other in all its dimensions. We all know that the volume, and therefore the weight of two similar bodies are proportional to the cubes of their dimensions. The cube of two is eight. Hence the large machine will have eight times the weight of the other. But surfaces are as the squares of the dimensions. The square of two is four. The heavier machine will therefore expose only four times the wing surface to the air, and so will have a distinct disadvantage in the ratio of efficiency to weight.

Mechanical principles show that the steam pressures which the engines would bear would be the same, and that the larger engine, though it would have more than four times the horse power of the other, would have less than eight times. The larger of the two machines would therefore be at a disadvantage, which could be overcome only by reducing the thickness of its parts, especially of its wings, to that of the other machine. Then we should lose in strength. It follows that the smaller the machine the greater its advantage, and the smallest possible flying machine will be the first one to be successful.

We see the principle of the cube exemplified in the animal kingdom. The agile flea, the nimble ant, the swift-footed greyhound, and the unwieldy elephant form a series of which the next term would be an animal tottering under its own weight, if able, to stand or move at all. The kingdom of flying animals shows a similar gradation. The most numerous fliers are little insects, and the rising series stops with the condor, which, though having much less weight than a man, is said to fly with difficulty when gorged with food.

We have also to consider the advantage which a muscle has over any motor yet discovered, in regard to its flexibility and the versatility of its application. It expands and contracts, pulls and pushes, in a way that no substance yet discovered can be made to do. It is also instantly responsive to a brain which cannot of itself act on external matter.

We may now see the kernel of the difficulty. If we had a metal so rigid, and at the same time so light, that a sheet of it twenty meters square and a millimeter thick would be as stiff as a board and would not weigh more than a ton, and, at the same time, so strong that a powerful engine could be built of it with little weight, we might hope for a flying machine that would carry a man. But as the case stands, the first successful flyer will be the handiwork of a watchmaker, and will carry nothing heavier than an insect. When this is constructed, we shall be able to see whether one a little larger is possible.

The cheapness of modern transportation is another element in the case frequently overlooked. I believe the principal part of the resistance which a limited express train meets is the resistance of the air. This would be as great for an airship as for a train. An important fraction of the cost of transporting goods from Chicago to London is that of getting them into vehicles, whether cars or ships, and getting them out again. The cost of sending a pair of shoes from a shop in New York to the residence of the wearer is, if I mistake not, much greater than the mere cost of transporting it across the Atlantic. I have shown that the construction of an aerial vehicle which could carry even a single man from place to place at pleasure requires the discovery of some new metal or some new force. Even with such a discovery, we could not expect one to do more than carry its owner.

Perhaps the main point I have tried to enforce in this paper is this—the very common and optimistic reply to objections, "We have seen many wonders, therefore nothing is impossible," is not a sound inference from experience when applied to a wonder long sought and never found. I have shown that the obvious and long-studied problems are not those that have been solved. The experience of the past leads us to believe that the progress of the twentieth century will be along lines that no one can anticipate, and will lead to results which, if a prophet could describe, might strike us as more surprising than the airship.

CHAPTER SIXTEEN
Modern Occultism

Simon Newcomb was a man of many interests. In the 1880s, his inquiries into what we would today call parapsychology resulted in his becoming president of the American Society for Psychical Research. Then as now, spiritualism and occultism attracted many believers, including such respected scientists as Oliver Lodge and William Crookes. It is probably fair to say that Newcomb was skeptical to begin with and as he dug into the subject his skepticism only increased. "Modern Occultism," which appeared only a few months before his death, is as sensible a discussion of the subject as one could hope to find, and is just as relevant today as when it was first published more than a century ago.

When eminent men of science announce discoveries of great interest it is an obvious general rule that their conclusions receive respectful consideration and, in the absence of strong reasons to the contrary, are accepted without serious question. But there is an exception to this rule so curious that it may well deserve our attention. Among the most important questions with which thought has been engaged are those of the possible modes of interaction between mind and mind. Coupled with this is the question of the direct action of mind upon matter, or of matter upon mind without physical agency. Ideas of this subject are older than civilization and arise so naturally that nothing but suggestion is necessary to implant them in the mind of the child. Discredited by the general trend of modern thought, the affirmative view has very generally been classed with superstition as belonging to a stage of intellectual development that the world has now left behind it. Belief in witchcraft vanished from the minds of civilized men more than two centuries ago, and with it disappeared the belief in every form of mental interaction otherwise than through the known organs of sense. But now men of eminence, whose opinion is entitled to the greatest respect, are informing us that the instincts of our ancestors did not err so greatly as we have supposed and that beliefs that our fathers called superstitious are well grounded in the regular order of nature. At least three scientific philosophers of the highest standing have placed themselves on record as accepting this view. Two of them, Sir Oliver Lodge and Professor William Barrett, have, during the past year, informed us that, not only is the direct transference of impressions from one mind to another a fact, but the spiritual world, which the thought of our time has been removing further and further from our everyday experience until it seemed likely to vanish from intellectual sight, is a reality knocking at our doors.

If these are truths, we can scarcely exaggerate their importance. Our most cherished aspirations and the consolations that religion offers to the dying and the bereaved are taken from the realm of sentiment and placed on the sure pedestal of science. A new view of mind is opened out, to the development of which we can set no limit. Accepting it, a system of conveying impressions from mind to mind at great distances, and of reading the secret thoughts of our fellows, seems more likely than it would have seemed a century ago that electricity would enable us to communicate with our antipodes. With such prospects opened out to us by scientific authorities so high, it certainly seems more appropriate that the skeptic, if such there be, should make known his reasons for the faith that is in him— perhaps we should say for his lack of faith—than that the doctrines should be treated as unworthy of attention.

A glance at the state of public opinion upon the subject will serve to guide the course of our thoughts. The class that fully accepts the views in question, notwithstanding its eminent respectability, is probably small in numbers. Between this class and those who entirely reject the views, as at least groundless, if not unworthy of consideration, there is an intermediate class holding that phenomena known as "occult" are exhibited that science has not yet satisfactorily explained. Their view has recently been happily stated by an able writer in the *Saturday Review*: "The existence of abnormal phenomena which science is only beginning to take notice of, a dim region of strange things which, even if they can be proved not to be supernatural, are at any rate outside the limits of organized experience," has been proved by the work of the Society for Psychical Research. "There are more things in heaven and earth than are dreamt of in your philosophy" has never ceased to express a feeling of the same general nature in the minds of intelligent men, and is at least one article of a creed always lending hope to the inquirer after the occult. This middle class, which thinks that there is something to learn in occultism, is certainly large, and perhaps makes up a majority of the intelligent community. It is to this class, as well as to that of believers, that the writer desires to address himself.

The personal element necessarily plays so large a part in any discussion of occultism that it may not be wholly out of place if the writer ventures on a brief statement of his own experience. The idea that the emotions of beloved relatives, sometimes at a great distance, might be agents in directing the various currents of feeling that run through the mind was imbibed in early childhood. Just how the idea originated he cannot say, but it is probably more common among children than we suspect. More than once, when hurrying home, he intently fixed his mind upon his mother with a strong desire that she should expect his coming, think about him, and prepare herself accordingly. But all these efforts proved failures. Another idea prevalent at a later period was that, by fixing the attention on someone sitting at a distance in front of you in church, you could move him to turn and look around him. But no systematic experiments in this direction were seriously attempted. When, in the early [1850s], the great wave of spiritualism, with its rappings, table-movings, and communications from the dead, was reaching its height, he naturally took an interest in the subject. But what little he could see of these performances seemed so silly as to prejudice him against the whole subject.

About 1858 an agent of prime importance in the history of spiritualism is worthy of being recalled. A warm discussion of the pretensions of certain mediums in the columns of the *Boston Courier* ended with the offer, by an anonymous writer (understood to be Professor Felton, afterward president of Harvard University), to pay a large reward to any mediums who would, in the presence of a committee to be named by himself, perform any of their pretended feats—move a table without touching it, read a paper in a closed envelope, or produce a rap the cause of which could not be traced. The offer was promptly accepted by the leader of the Boston spiritualists, and several of the most famous mediums were brought from different parts of the country. The committee was three in number. At its head was Professor Louis Agassiz, and his coadjutors were two eminent scientific men of Cambridge. The séances were held in the room of a Boston hotel. The result was a failure so complete that the professors felt humiliated to sit hour after hour and see nothing to enliven the proceedings. Some cabinet feats of tying and untying were attempted, but nothing was done in this line except very elementary tricks of legerdemain. The mediums could assign no better reason for their failure than the contempt of the spirits for men who disbelieved in their existence. A large measure of abuse was heaped upon the committee by the spiritualists, but no argument better than this was adduced in explanation of their failure.

After this the general attitude of the writer toward the subject was this: "I have no time to engage in the search after wonders. But tell me in any special case when I can go to a séance with any reasonable chance of seeing something out of the usual order of nature, and I will avail myself of the opportunity with alacrity." What has especially struck him ever since has been the absence of any such opportunity. When he was told of wonderful phenomena, and inquired as to details, the stories were always about things that had happened long before. An inquiry where a medium of special power could be found elicited no answer but that her whereabouts was unknown and she had probably left the city.

But after many years of waiting, an opportunity was at last presented. The most wonderful performer yet seen came to Washington, and her feats were vouched for by a party of intelligent gentlemen who had been invited to a private exhibition of her powers. She was a Miss Lulu Hirst, of Georgia. It must be said that spiritualism, as well as any other theory, was ignored by her; but this was a minor matter, as the feats were of the same kind as those essayed by the professional spiritualists. A day or two later arrangements were made for another series of tests, in which the writer took part. Without going into details, which were published fully at the time, it will suffice to remark in the present connection that nothing was shown but what was obviously produced by the efforts of a muscular and dexterous young woman. She was quite frank and honest, without pretences to be investigated or trickery to be exposed. Every surprising element in the narrative proved to be based on imperfections of observation and misconception of what was seen. Only one feature was needed to complete the picture. When the public performance of the "wonder-girl" came off, the press reporters were, of course, present, and their accounts of her feats as narrated in the journals rivaled or outdid the performances of the most celebrated mediums.

After the English Society for Psychical Research was organized by a body of men eminent in various fields of thought and action, the past failures of the writer did not prevent his taking part in the formation of an American society of the same kind, of which he had the honor to be elected the first president. Two years of experiment, study, and reading confirmed his ideas on the subject, but he remained for some time longer in occasional communication and cooperation with Dr. Hodgson, a well-known member of the English Society, then resident in Boston. He now invites the courteous consideration of the reader to the views of the subject that he has reached after a half-century of occasional study coupled with reading the best he could find in support of occultism.

We may approach the heart of our subject in the easiest way by recalling two lines of research in which Sir William Crookes took a prominent part. The name of this eminent investigator has become a household word in science from his discovery that a singular radiance may be produced at the cathode of a vacuum tube through which an electric current is passing. He also observed curious phenomena of motion among material objects in his laboratory for which he could not assign any physical cause. Several years elapsed after these discoveries before either of them seemed destined to develop into an important branch of science. Then the one first mentioned suddenly assumed importance.

In 1895 Professor Röntgen made the astounding discovery that certain rays from a Crookes tube were capable of passing through opaque substances and imprinting themselves upon a photographic plate beyond. About the same time it was shown by Becquerel that rays of similar properties, but different in kind, could be produced from uranium. All the physical laboratories of the world were at once actively engaged in testing these discoveries and following up the lines of research they suggested. The result was the discovery of radium and the development of a new branch of physics—radioactivity, which has gone on expanding until it bids fair to revolutionize our views of matter, ether, and their relations. Works on radioactivity are multiplying, and physicists are looking for new theories of light and electricity that are to grow out of this field of research.

With this outcome in mind, let us trace up the lines of the other observation. More than ten years before Röntgen's work the Society for Psychical Research had been organized. The special purpose was the critical investigation of occult phenomena in general, especially those that seemed to show the passage of impressions from mind to mind without material agency. A discovery that seemed to inaugurate a revolution in science of mind was soon announced in the form of an experiment equally remarkable for its simplicity and its importance. A blindfolded person, called a "percipient," was seated at a table with pencil in hand and paper before him, while his senses, especially those of sight and touch, were protected so far as possible from the action of all external agencies. His mind was to be quite free from all prepossession and his will to be reduced as nearly as possible to a state of quiescence. The only action allowed was that of drawing geometrical figures on the paper quite at random, without intent to produce any special forms. Behind him, but not in contact or communication, was seated an "agent" with a miscellaneous collection of geometrical figures. While the agent concentrated his vision and attention as intensely as possible upon one of these, the percipient was instructed to allow his pencil to move on the paper without any prejudice in favor of any special form of motion. The process was repeated with one figure after another. When the drawings of the percipient were compared with the originals, a resemblance was found sufficient to show an undoubted relation between the reproduced figures and those on which the attention of the agent had been fixed.

The experiments were not confined to geometric forms. Others were devised with the common object of showing that the random actions of one mind were affected by the action of another mind in its neighborhood, without the use of words or signs. When the agent drew cards from a pack one by one, and at each drawing the percipient named a card at random, it was found that the proportion of correct guesses was much greater than it should have been as the result of chance, which would, of course, be 1 out of 52.

In one point these experiments had a great advantage over those of the physicists. Crookes tubes and other apparatus required for experiments in radioactivity demand so much care and expense in their production that their use is confined to professional workers in physical laboratories. But the apparatus necessary to the demonstration of thought-transference abounds in every household. Men, women, paper, pencils, tables, screens, handkerchiefs for blindfolding, and cards make up a fairly complete list of essentials. The results to be ultimately expected from the experiments transcend in practical importance all that we can expect from the development of radioactivity. Such being the case, the natural anticipation was that thought-transference would become a branch of experimental psychology, the laws of which would form an important chapter in every treatise on this subject, and that apparatus for showing it would be as well known in every psychological laboratory as that for experimenting in X-rays is in every physical laboratory.

Twenty-five years have elapsed since the announcement, and what has been the outcome? Scientifically, nothing at all. The science of psychology has been behind few others in the extent of its development since the experiments described were begun. But if thought-transference is seriously treated in any treatise on this science the writer has not noticed it. The reason is not far to seek. No result relating to thought-transference has yet been reached that belongs to the realm of science. Science properly so-called comprises the statement of laws or general facts. No collection of isolated events, however large it may be, forms a part of it. Radioactivity is a science because it is a general fact that everyone can verify that, if you organize a certain system of experiments, you can take a photograph through many opaque substances. That coal will burn when brought into contact with fire is a proposition belonging to the same domain. But if we could only say that someone in England had at some time made coal burn, then, a few years later, someone in Russia, then someone in America, and so on, such facts, though they mounted into the hundreds or the thousands, would not establish the law that coal was combustible, and therefore would not belong to science. The question of how the supposed burning came about in the special cases cited might be interesting, yet the process of investigation would be difficult if no careful experimenter were ever able to bring the combustion about.

So with thought-transference. In order that a scientific conclusion as to its reality may be reached, it is necessary to show under what conditions it takes place. The Psychical Society tried to determine, by a repetition of the experiments under various conditions, whether the action of the agent upon the percipient would pass through a screen, and how it varied with the conditions. When these questions could be answered, the first step would be taken toward placing the subject upon a scientific basis. But no result could ever be reached that was general in form. The nearest approach to a general proposition that could be formulated from all the experiments was: If you make the experiment you may possibly see what seems to show thought-transference, and you may not. The probability of success cannot be stated because we have no record of the failures, the number of which defies estimation. I have tried to learn whether during the past ten years the Psychical Society has done anything toward elucidating the subject. But nothing bearing on the case is found in its recent published proceedings. Would it be altogether unfair to put the conclusion in the form: Possibly you may succeed, but the more pains you take to avoid all sources of error, the less likely success will be?

During the past fifteen years interest has been transferred from thought-transference to telepathy. The question of how, if an impression cannot be conveyed through a space of a few feet, it can yet dart from one city to another is one that, how strongly soever it may present itself, may rest in abeyance while we inquire about the seeming facts. These, as found in the fine volumes *Phantasms of the Living*, by Gurney and Myers, and in the publications of the Psychical Society, are too numerous to be summarized. But a typical example that will answer our present purpose is easy to give. A person is struck by a sudden hallucination, or has a vision or dream of a friend or relative, generally in distress. This impression is so vivid that some anxiety may be felt lest it correspond to a reality. Next morning, or as soon as the mail or telegraph can bring the news, it is learned that the friend or relative has either died at the time of the vision or has suffered some violent emotion. Great pains were taken to verify the authenticity of stories of this kind, and none were accepted unless deemed "veridical." Taking the hundreds of coincidences as they stand, and regarding each narrative as complete in itself, the conclusion that there must have been some causal connection between the distant event or emotion and the vision looks unavoidable. But may it not be that causes already known are sufficient to account for the supposed coincidences without introducing telepathy or any other abnormal agency? If such is the case, then the hypothesis of telepathy is purely gratuitous and uncalled for, on the general principle that we never attribute events to new and unknown causes when we see that they are the natural results of known conditions. This is especially the case when the new causes deduced are so improbable and so far outside the line of our general experience as telepathy must be. The strongest believer in this agency must admit that its acceptance is not without difficulty. Everyone who sleeps in London is surrounded by several millions of minds within a radius of three or four miles. Among these are hundreds in a state of violent action or emotion. Scores are constantly in the throes of death. How do the inhabitants of London sleep on undisturbed by the spiritual tumult? How is it that in the ordinary experience of life one person cannot divine the most intense feeling of another, even though he be near or dear, except by sight, touch, or hearing? So far as the writer is aware, the advocates of telepathy have evaded rather than grappled with these difficulties.

The question we shall now consider is whether there are not known causes at play that we should naturally expect to result in phenomena that seem to indicate telepathy. Those that I shall adduce are not all of one kind, but are made up of complex elements, each of which is familiarly known to all who carefully think and observe. First to be mentioned is the element of truth. Then will come the omission of important features from the narrative. I believe that Bacon remarked that men score only the hits, and ignore the misses. We also have unconscious exaggeration; the faculty of remembering what is striking and forgetting what is not; illusions of sense, mistakes of memory; the impressions left by dreams; and, finally, deceit and trickery, whether intentional or unconscious. Before reaching a conclusion we must inquire as to what we should naturally expect as the combined result of these agencies in the regular course of experience.

As to the first: error finds support in so entwining itself with truth that it is difficult to separate the two. Double personality, hypnotism, and especially the action of one mind on another by hypnotic suggestion, have been confused with telepathy through a supposed power of the operator to influence the will of his subject at a distance. The mystery that has very generally enveloped the subject of "animal magnetism" is so fertile in vague theories of abnormality that now, when the whole subject is placed on a scientific basis, the elimination of traditional and baseless ideas is by no means an easy task. The belief that a hypnotic operator influences his subject by telepathy is widely diffused through all classes of the community except professional psychologists. The latter are, I believe, practically unanimous in holding that no influence is exerted on the subject except through the medium of the senses, and that, if the subject is to act in a certain way in the absence of the operator, the latter must make known in advance the time and nature of the expected action. I am aware that Richet and perhaps other operators have found cases that seem telepathic, but a critical reading of their evidence shows it to be wholly inconclusive.

A course of events may appear ever so wonderful and incomprehensible by well-known agencies by mere omission, without deviating from the truth in any particular. I once examined an interesting case of this kind at the request of Dr. Hodgson. A naval ship had been wrecked in a storm off Cape Hatteras some years before, and most of those on board, including the captain, had perished. Before she sailed on her voyage one of her officers was seized with so strong and persistent a presentiment that the ship would be lost that he formally requested to be detached from her. This being refused, he left his post of duty and was tried by court martial for desertion. Dr. Hodgson desired me to see whether this story could be verified by the official records.

This was easily done, and the narrative was found to be substantially correct so far as it went. But it omitted to state that the officer had exhibited symptoms of mental aberration before his presentiment, that the latter was only one of a great number of wild fears that he had expressed to various parties, including his superior officer, and that several months elapsed after this before the ship sailed on her fateful voyage, she having in the meantime made several trips on the coast. When thus completed the story became altogether commonplace.

A coincidence between an emotion experienced by a distant person and the impression of that emotion in another at a distance can indicate a causal relation only when the coincidence is real and the impression unusual. In establishing the facts there is wide ground for error. We are all subject to errors of memory, especially if we have to state the exact time and circumstance of an act or impression. Probably few of us could tell all that we did the day before yesterday, hour by hour, without either some erroneous statement, the omission of some act, or the introduction of an event that belonged to a different day. The longer the time that elapses, the greater the liability to error. Writers on telepathy take too little account of these errors of memory. In the vast majority of cases the correction cannot be made, and the error goes on record as truth, when it becomes the basis for some remarkable coincidence. When this is not the case it passes into oblivion. If we set a net for errors that we cannot distinguish from truth, how shall we know that our catch is anything but error? It is only by having some independent test of the accuracy of a remembered event that we can be sure of its correctness. A written and dated document, if genuine, would always suffice for this purpose. But such support is almost if not quite universally wanting in the narratives of wonderful coincidences.

I only recall a single case in which the correctness of a telepathic narrative was tested by independent and conclusive authority. In the *Nineteenth Century* for July 1884, an article, "Apparitions," by Edmund Gurney and Frederic W. H. Myers, appeared that was justly regarded as affording, the most indisputable evidence ever adduced for the reappearance of a dead person. Sir Edmund Hornby, a judge of the Consular Court at Shanghai, had been visited during the night by a reporter desiring a copy of a decision that he was to deliver on the following morning. He rose from his bed, dictated what he had to say, and dismissed the reporter with a rebuke for having disturbed him. Next morning, on going to court, he was astounded by learning that the reporter, with whom he was well acquainted, had died suddenly during the night. Inquiring after the hour of the demise he found it to coincide with that of the nightly visitation. The authors also informed us in the article that the story was confirmed by Lady Hornby, who was mentioned in it and was cognizant of the circumstances.

This narrative was almost unique in that it admitted of verification. When it reached Shanghai it met the eyes of some acquainted with the actual facts. These were made known in another publication and showed that several months must have elapsed between the reporter's death and the judge's vision. The latter was only a vivid dream about a dead person. When the case was brought to the judge's attention he did not deny the new version and could only say he had supposed the facts to be as he had narrated them.

I cite this incident not merely to show how the most conclusive case of telepathy ever brought to light was invalidated when the facts were made known but to elucidate the further fact that a wonderful story may lose the element of surprise by quite natural and easily admitted additions and explanations. All the interest of such stories depends upon the element of wonder.

The looker-on feels most delight
Who least perceives the juggler's sleight.

It is positively humiliating to allow an amateur juggler to explain his extraordinary tricks. It humiliates one that he did not himself see how the thing was done. Why should we hesitate to ascribe any number of seemingly supernatural occurrences to the innumerable blunders that we know nearly every one of us is making in memory every day?

The statistical one-sidedness of all evidence in favor of telepathy, apparitions, and other forms of supernormal mental action must be considered, and so far as possible corrected, before any conclusion can be reached. The principle involved and the ease with which we may reach a false conclusion may be illustrated by a very simple example. If a bag of corn contains a million normal grains and a single black one, the probability that a grain drawn at random from the bag would be the black one is so minute that we should justly regard the drawing as practically impossible in all the ordinary affairs of life. If a blindfolded boy, dipping his hand into the bag, drew the black grain on the first trial, we should justly claim that there was some unfairness in the proceeding, or, if we wish to deal in mystery, some attraction between his hand and the black grain. If on a thousand trials of this kind the black grain was drawn several times our suspicion would ripen into practical certainty. And yet, if every inhabitant of Great Britain made such a trial, it is practically certain that there would be about thirty drawings of the black grain without abnormality. In fact, did such drawings number only twenty, the suspicion would be on the other side. We should be sure of some defect in the enumeration or of some instinct toward evading the black grain. The whole question turns on the number of unrecorded failures.

Through inquiries made under the auspices of the Psychical Society it would seem that about one person in every ten is more or less subject to hallucinations of some kind. Probably a large majority of people have occasional dreams so vivid that in Great Britain alone there must occur annually many millions of cases in which people, during their waking or dreaming hours, see before them images of distant relatives or friends. If, as may well be the case, the chances are millions to one against the illusion coinciding with the death or distress of the person seen, we should still have in all probability many such cases in a year. Thus, when the eminent members of the Society instituted their inquiries for such cases, it might have been predicted in advance that, without any bias whatever, they would have been discovered by the hundred.

But the concession of exactness is one of great improbability. Visions and dreams are in all ordinary cases dropped from the mind and speedily forgotten. But let one be connected in any way with a death or other moving event, and the memory, instead of being effaced, grows in the mind, month after month. The event associated with the vision may have occurred days or weeks before or after it, but the general tendency will be to bring them into coincidence and weave them into a story, as we have seen in the case already quoted.

The following case, cited by Mr. Beckles Willson in his recent work, *Occultism and Common Sense*, may be chosen for study because it is among the most remarkable of its kind. A traveler in a railway carriage is quoted:

> One week ago last Tuesday, at eleven o'clock at night, my wife, who had just retired to bed upstairs, called out to me, "Arthur! Arthur!" in a tone of alarm. I sprang up and ran upstairs to see what was the matter. The servants had all gone to bed. "Arthur," said my wife, "I've just seen mother," and she began to cry. "Why," I said, "why, your mother is in

Scarborough." "I know," she said; "but she appeared before me just there" (pointing to the foot of the bed) "two minutes ago as plainly as you do." Well, the next morning there was a telegram on the breakfast table— "Mother died at eleven last night." Now, how do you account for it?

I will try to answer this question. I would not be at all surprised, could the facts be made known, if the wife had said something of the kind to her husband every day or night for a week, especially if the mother were known to be very ill. If any night had been missed, I would not be surprised if it were the fateful Tuesday. Then the problem would have been reversed, and we should have had to explain why it was that the vision failed on the night of the death. The memory of the narrator had more than a week in which to cultivate the wonder. The quotation, it will be noticed, purports to be verbatim, though, from what the author says, many years had probably elapsed. During this time the wonder, as it came from the lips of the original speaker, had ample time to develop still further in the mind of the narrator. What limit can we set to its possible growth, first in one mind and then in another? I cannot but feel that the more experience the reader has had in observing this form of growth, the less he will be inclined to set any limit to it.

Considering the natural processes of adaptation and exaggeration, from which no mind is so well disciplined as to be absolutely free, we conclude that the annual number of seeming but groundless telepathic phenomena in Great Britain alone is probably to be counted by thousands. The volumes of *Phantasms of the Living* might be continued annually without end could all the cases be discovered. The few hundred cases published are actually fewer than what we should expect as the result of known conditions. There is therefore no proof of telepathy in any of the wonders narrated in these volumes, and in the publications of the Psychical Society.

We have considered the evidence for the various forms of telepathy with some fullness because the theory is, in form at least, a scientific one, and the evidence admits of being treated by the established methods of logical inference. But telepathy is only the beginning of the wonders collected by modern inquirers into the occult, who find so many phenomena unexplainable, even by this agency, that they regard the latter as only a first step in the science they are trying to construct. Our conclusion from all these supposed phenomena are so much matters of individual judgment, not admitting of being readily reduced to first principles, that they must be disposed of quite briefly. The belief in specially gifted persons—doers of miracles and practitioners of witchcraft—was once almost universal. Our modern students of occultism have revived what seems very like these discarded beliefs, though the word "witchcraft" is no longer used to express the abnormal powers in question. These powers are not merely those possessed by men in general and heightened in degree, like the faculty of the lightning calculator or the muscular dexterity of the acrobat; but they are powers of which men in general are absolutely devoid. Examples of them are levitation, clairvoyance, ability to make oneself seen in distant places, to move objects without touching them, to put one's head into the fire or walk over burning coals without injury, and as many others as ingenuity can suggest. Men are still living who testify to having seen a medium rise in the air and waft himself around a room, or disappear through a window.

Now, if we admit the existence of gifted individuals having such abnormal powers as these, why not equally admit the existence of men having the faculty of seeing, or thinking they remember having seen, the non-existent? The latter certainly seems much easier to suppose than does the former. It is a familiar fact of physiological optics that, in a faint light, if the eyes are fixed upon an object, the latter gradually becomes clouded and finally disappears entirely. Then it requires only a little heightening of a not unusual imagination to believe that, if the object that disappeared was a man, he wafted himself through the air and went out of the window.

What are we to say of the performances of mediums, tiers and untiers of hands, table-rappers, slate-writers, cabinet-workers, materializers, and the whole class of performers to which they belong? May we not adduce the general principle that similar phenomena are to be attributed to similar causes? These performances are quite similar to those of legerdemain, which we may witness for a few shillings in broad daylight at any exhibition of the juggler's art. The principal point of difference is that they are less wonderful and, being generally seen in a faint light, give much greater opportunity for trickery than do those of the professional operators on the stage. Is it logical to attribute them to occult causes when we regard the professional performers as mere mystifiers? This question seems to the writer to answer itself.

I have not considered the supernatural knowledge supposed to be possessed by the "trance-medium" because the data for reaching any conclusion on the subject are too vague to admit of precise statements. The careful examination of Mrs. Piper made by the Psychical Society several years ago is unique in that the proceedings were reported stenographically. A few of her expressions did seem to show supernatural knowledge of, or impression by, facts with which she could not have been acquainted by any natural process. But the relation was wanting in that definiteness on which alone a positive conclusion could be based. The balancing of the probabilities on the two sides can well be made by everyone for himself.

In reaching a general conclusion upon all the evidence for the occult I would lay special stress on a feature already mentioned in narrating my personal experience. Almost all the narratives I have seen or heard relate to experiences of years previous, and scarcely ever to the present, so that the wonder has plenty of time to grow in the memory. The latest work on occultism with which I am acquainted is that of Mr. Willson, already cited. Turning over its leaves I fail to find any occurrence, in England at least, of later date than 1896, 12 years before publication. There are a few dubious-looking reports from other countries of a little later date than this, but nothing of the present time. Except the trance-mediums and fortune-tellers, who still ply their trade, and an occasional "materializer," the writer has heard nothing of mediumistic performances for ten or even twenty years. Why do

> *Peor and Baalim*
> *Forsake their temples dim?*

Is it not because in the course of years a wonder grows in the memory, like an oak from an acorn? The writer fails to see how a sane review of the whole subject can lead to any other conclusion than that occultism has no other basis than imperfect knowledge of the conditions, or how a wide survey of the field can leave any room for mystery.

We live in a world where in every country there are millions of people subject to illusions too numerous to be even classified. They arise from dreams, visions, errors of memory that can rarely be detected, and mistakes to which all men are liable. It is unavoidable that when any of these illusory phenomena are associated with a moving event at a distance, there will be an apparent coincidence that will seem more wonderful every time it is recalled in memory. There is no limit to devices by which ingenuity may make us see what is unreal. Every country has ingenious

139

men by the thousands, and if a willingness to deceive overtly characterizes only a small fraction of them, that fraction may form so large a number of individuals, always ready to mystify the looker-on, that the result will be unnumbered phenomena apparently proving the various theories associated with occultism and spiritualism. Nothing has been brought out by the researches of the Psychical Society and its able collaborators except what we should expect to find in the ordinary course of nature. The seeming wonders—and they are plentiful—are at best of the same class as the wonder when a dozen drawers of the black grain of corn out of a million are presented to us. We are asked to admit an attraction between their hands and the black grain. The proof is conclusive enough until we remember that this dozen is only a selection out of millions, the rest of whom have not drawn the black grain. The records do not tell us, and never can tell us, about the uncounted millions of people who have forgotten that they ever had a vision or any illusion, or who, having such, did not find it associated with any notable occurrence. Count them all in, and nothing is left on which to base any theory of occultism.

CHAPTER SEVENTEEN

Can We Make It Rain?

The problem of making rain by artificial means was another area where Newcomb sought to apply the scientific method to a problem with social and economic consequences. Here he was on firmer ground than in the case of heavier-than-air flight. Trying to create precipitation during a drought by firing off ordnance or explosives came into the vogue during the late nineteenth century and revived in popularity during the Dust Bowl years. It didn't work, as Newcomb points out. From the mid-twentieth century onward, attempts have been made to "seed" rain clouds using silver iodide or dry ice, which provide nuclei upon which raindrops can form where they otherwise might not. Such methods have met with only limited success. Rainmaking aside, it is hard to dispute Newcomb's closing observation that without proper scientific and technical advice, politicians are all too prone to waste large amounts of money on projects of doubtful (or even no) utility.

T o the uncritical observer the possible achievements of invention and discovery seem boundless. Half a century ago no idea could have appeared more visionary than that of holding communication in a few seconds of time with our fellows in Australia, or having a talk going on *viva voce* between a man in Washington and another in Boston. The actual attainment of these results has naturally given rise to the belief that the word "impossible" has disappeared from our vocabulary. To every demonstration that a result cannot be reached the answer is, "Did not one Lardner, some sixty years ago, demonstrate that a steamship could not cross the Atlantic?" If we say that for every actual discovery there are a thousand visionary projects, we are told that, after all, any given project may be the one out of the thousand.

In a certain way these hopeful anticipations are justified. We cannot set any limit either to the discovery of new laws of nature or to the ingenious combination of devices to attain results which now look impossible. The science of today suggests a boundless field of possibilities. It demonstrates that the heat which the sun radiates upon the earth in a single day would suffice to drive all the steamships now on the ocean and run all the machinery on the land for a thousand years. The only difficulty is how to concentrate and utilize this wasted energy. From the standpoint of exact science aerial navigation is a very simple matter. We have only to find the proper combination of such elements as weight, power, and mechanical force. Whenever Mr. Maxim can make an engine strong and light enough, and sails large, strong, and light enough, and devise the machinery required to connect the sails and engine, he will fly. Science has nothing but encouraging words for his project, so far as general principles are concerned. Such being the case, I am not going to maintain that we can never make it rain.

But I do maintain two propositions. If we are ever going to make it rain, or produce any other result hitherto unattainable, we must employ adequate means. And if any proposed means or agency is already familiar to science, we may be able to decide beforehand whether it is adequate. Let us grant that out of a thousand seemingly visionary projects one is really sound. Must we try the entire thousand to find the one? By no means. The chances are that nine hundred of them will involve no agency that is not already fully understood, and may, therefore, be set aside without even being tried. To this class belongs the project of producing rain by sound. As I write, the daily journals are announcing the brilliant success of experiments in this direction; yet I unhesitatingly maintain that sound cannot make rain, and propose to adduce all necessary proof of my thesis. The nature of sound is fully understood, and so are the conditions under which the aqueous vapor in the atmosphere may be condensed. Let us see how the case stands.

A room of average size, at ordinary temperature and under usual conditions, contains about a quart of water in the form of invisible vapor. The whole atmosphere is impregnated with vapor in about the same proportion. We must, however, distinguish between this invisible vapor and the clouds or other visible masses to which the same term is often applied. The distinction may be very clearly seen by watching the steam coming from the spout of a boiling kettle. Immediately at the spout the escaping steam is transparent and invisible; an inch or two away a white cloud is formed, which we commonly call steam, and which is seen belching out to a distance of one or more feet, and perhaps filling a considerable space around the kettle; at a still greater distance this cloud gradually disappears. Properly speaking, the visible cloud is not vapor or steam at all, but minute particles or drops of water in a liquid state. The transparent vapor at the mouth of the kettle is the true vapor of water, which is condensed into liquid drops by cooling; but after being diffused through the air these drops evaporate and again become true vapor. Clouds, then, are not formed of true vapor, but consist of impalpable particles of liquid water floating or suspended in the air.

But we all know that clouds do not always fall as rain. In order that rain may fall the impalpable particles of water which form the cloud must collect into sensible drops large enough to fall to the earth. Two steps are therefore necessary to the formation of rain: the transparent aqueous vapor in the air must be condensed into clouds, and the material of the clouds must agglomerate into raindrops.

No physical fact is better established than that, under the conditions which prevail in the atmosphere, the aqueous vapor of the air cannot be condensed into clouds except by cooling. It is true that in our laboratories it can be condensed by compression. But, for reasons which I need not explain, condensation by compression cannot take place in the air. The cooling which results in the formation of clouds and rain may come in two ways. Rains which last for several hours or days are generally produced by the intermixture of currents of air of different temperatures. A current of cold air meeting a current of warm, moist air in its course may condense a considerable portion of the moisture into clouds and rain, and this condensation will go on as long as the currents continue to meet. In a hot spring day a mass of air which has been warmed by the sun, and moistened by evaporation near the surface of the earth, may rise up and cool by expansion to near the freezing-point. The resulting condensation of the moisture may then produce a shower or thunder-squall. But the formation of clouds in a clear sky without motion of the air or change in the temperature of the vapor is simply impossible. We know by abundant experiments that a mass of true aqueous vapor will never condense into clouds or drops so long as its temperature and the pressure of the air upon it remain unchanged.

Now let us consider sound as an agent for changing the state of things in the air. It is one of the commonest and simplest agencies in the world, which we can experiment upon without difficulty. It is purely mechanical in its action. When a bomb explodes, a certain quantity of gas, say five or six cubic yards, is suddenly produced. It pushes aside and compresses the surrounding air in all directions, and this motion and compression are transmitted from one portion of the air to another. The amount of motion diminishes as the square of the distance; a simple calculation shows that at a quarter of a mile from the point of explosion it would not be one ten-thousandth of an inch. The condensation is only momentary; it may last the hundredth or the thousandth of a second, according to the suddenness and violence of the explosion; then elasticity restores the air to its original condition and everything is just as it was before the explosion. A thousand detonations can produce no more effect upon the air, or upon the watery vapor in it, than a thousand rebounds of a small boy's rubber ball would produce upon a stone wall. So far as the compression of the air could produce even a momentary effect, it would be to prevent rather than to cause condensation of its vapor, because it is productive of heat, which produces evaporation, not condensation.

The popular notion that sound may produce rain is founded principally upon the supposed fact that great battles have been followed by heavy rains. This notion, I believe, is not confirmed by statistics; but, whether it is or not, we can say with confidence that it was not the sound of the cannon that produced the rain. That sound as a physical factor is quite insignificant would be evident were it not for our fallacious way of measuring it. The human ear is an instrument of wonderful delicacy, and when its tympanum is agitated by a sound we call it a "concussion" when, in fact, all that takes place is a sudden motion back and forth of a tenth, a hundredth, or a thousandth of an inch, accompanied by a slight momentary condensation. After these motions are completed the air is exactly in the same condition as it was before; it is neither hotter nor colder; no current has been produced, no moisture added.

If the reader is not satisfied with this explanation, he can try a very simple experiment which ought to be conclusive. If he will explode a grain of dynamite, the concussion within a foot of the point of explosion will be greater than that which can be produced by the most powerful bomb at a distance of a quarter of a mile. In fact, if the latter can condense vapor a quarter of a mile away, then anybody can condense vapor in a room by slapping his hands. Let us, therefore, go to work slapping our hands, and see how long we must continue before a cloud begins to form.

What we have just said applies principally to the condensation of invisible vapor. It may be asked whether, if clouds are already formed, something may not be done to accelerate their condensation into raindrops large enough to fall to the ground. This also may be the subject of experiment. Let us stand in the steam escaping from a kettle and slap our hands. We shall see whether the steam condenses into drops. I am sure the experiment will be a failure; and no other conclusion is possible than that the production of rain by sound or explosions is out of the question.

It must, however, be added that the laws under which the impalpable particles of water in clouds agglomerate into drops of rain are not yet understood, and that opinions differ on this subject. Experiments to decide the question are needed, and it is to be hoped that the Weather Bureau will undertake them. For anything we know to the contrary, the agglomeration may be facilitated by smoke in the air. If it be really true that rains have been produced by great battles, we may say with confidence that they were produced by the smoke from the burning powder rising into the clouds and forming nuclei for the agglomeration into drops, and not by the mere explosion. If this be the case, if it was the smoke and not the sound that brought the rain, then by burning gunpowder and dynamite we are acting much like Charles Lamb's Chinamen who practised the burning of their houses for several centuries before finding out that there was any cheaper way of securing the coveted delicacy of roast pig.

But how, it may be asked, shall we deal with the fact that Mr. Dyrenforth's recent explosions of bombs under a clear sky in Texas were followed in a few hours, or a day or two, by rains in a region where rain was almost unknown? I know too little about the fact, if such it be, to do more than ask questions about it suggested by well-known scientific truths. If there is any scientific result which we can accept with confidence, it is that ten seconds after the sound of the last bomb died away, silence resumed her sway. From that moment everything in the air—humidity, temperature, pressure, and motion—was exactly the same as if no bomb had been fired. Now, what went on during the hours that elapsed between the sound of the last bomb and the falling of the first drop of rain? Did the aqueous vapor already in the surrounding air slowly condense into clouds and raindrops in defiance of physical laws? If not, the hours must have been occupied by the passage of a mass of thousands of cubic miles of warm, moist air coming from some other region to which the sound could not have extended. Or was Jupiter Pluvius awakened by the sound after two thousand years of slumber, and did the laws of nature become silent at his command? When we transcend what is scientifically possible, all suppositions are admissible; and we leave the reader to take his choice between these and any others he may choose to invent.

One word in justification of the confidence with which I have cited established physical laws. It is very generally supposed that most great advances in applied science are made by rejecting or disproving the results reached by one's predecessors. Nothing could be farther from the truth. As Huxley has truly said, the army of science has never retreated from a position once gained. Men like Ohm and Maxwell have reduced electricity to a mathematical science, and it is by accepting, mastering, and applying the laws of electric currents which they discovered and expounded that the electric light, electric railway, and all other applications of electricity have been developed. It is by applying and utilizing the laws of heat, force, and vapor laid down by such men as Carnot and Regnault that we now cross the Atlantic in six days. These same laws govern the condensation of vapor in the atmosphere; and I say with confidence that if we ever do learn to make it rain, it will be by accepting and applying them, and not by ignoring or trying to repeal them.

How much the indisposition of our government to secure expert scientific evidence may cost it is strikingly shown by a recent example. It expended several million dollars on a tunnel and water-works for the city of Washington, and then abandoned the whole work. Had the project been submitted to a commission of geologists, the fact that the rock-bed under the District of Columbia would not stand the continued action of water would have been immediately reported, and all the money expended would have been saved. The fact is that there is very little to excite popular interest in the advance of exact science. Investigators are generally quiet, unimpressive men, rather diffident, and wholly wanting in the art of interesting the public in their work. It is safe to say that neither Lavoisier, Galvani, Ohm, Regnault, nor Maxwell could have gotten the smallest appropriation through Congress to help make discoveries which are now the pride of our century. They all dealt in facts and conclusions quite devoid of that grandeur which renders so captivating the project of attacking the rains in their aerial stronghold with dynamite bombs.

CHAPTER EIGHTEEN

What Is a Liberal Education?

In the following brief paper, first published in Science *on 11 April 1884, Newcomb argues (as he would later and at greater length in "The Relation of Scientific Method to Social Progress") that the scientific method should be central to a liberal education. Such a view no doubt struck many of Newcomb's contemporaries as heretical in an age when the classics were widely viewed as the indispensable core of the university curriculum. Equally modern is his suggestion that it is the education of today's young people that will make the most difference to the state of the world fifty years hence. The challenges facing us are, if anything, even greater than those which confronted Newcomb's world, and the answers still seem to lie in what Newcomb here calls "organized common sense," making this essay a fitting one with which to close the book.*

I do not intend, in the present paper, to enter upon the disputed question between the advocates of classical culture on the one hand, and those of scientific training on the other; because it seems to me that the line on which the two parties divide is not that which really divides the thought of the day. If we look closely into the case, we shall see that the objects of a higher education may be divided into three classes, instead of the two familiar ones of liberal and professional. In fact, what we commonly call a liberal education should, I think, have two separate objects. With the idea of a professional education we are all familiar: it is that which enables the possessor to pursue with advantage some wealth-producing specialty. Although, in accordance with well-known economic principles, it is designed to make the individual useful to his fellow men, the ultimate object in view is the gaining of a livelihood by the individual himself. On the other hand, the object had in view in what is commonly known as culture is not the mere gaining of a livelihood, but the acquisition of those ideas, and the training of those powers, which conduce to the happiness of the individual. From this point of view, culture may be considered an end unto itself.

The third object which we have to consider is only beginning to receive recognition in the eyes of the public. It is the general usefulness of the individual, not merely to himself and to those with whom he stands in business relations, but to society at large. Modern thought and investigation lead to the conclusion that man himself, the institutions under which he lives, and the conditions which surround him are subject to slow, progressive changes; and that it depends very largely on the policy of each generation of mankind whether these changes shall be in the way of improvement or retrogression. During the next fifty years all of us will have passed from the stage of active life, and the course of events will be very largely directed by men who are still unborn. The happiness of those men is, from the widest philanthropic point of view, just as important as the happiness of those who now inhabit the earth; and, in the light of modern science, we now see that that happiness depends very largely upon our own actions. We thus have opened out to us an interest and a field of solicitude in which we need the best thought of the time. The question is, what form of education and training will best fit the now rising generation for the duty of improving the condition of the generation to follow it?

Let it be understood that we are now speaking not of the education of the masses, but of that higher education which is necessarily confined to a small minority. So far as I am aware, that fraction of the male population which receives a college education is not far from one per cent. To that comparatively small body we must look for the power which is to direct the society of the future, and by their acts to promote the well- or ill-being of the coming generation. Our duty to that generation is to so use and train this select body as to be of most benefit to the men of the future. What is the training required? I reply by saying that I know nothing better for this end than a wide and liberal training in the scientific spirit and the scientific method. The technicalities of science are not the first object; and, so far as they are introduced, it is only as media through which we may imbue the mind with certain general and abstract ideas. If called upon to define the scientific spirit, I should say that it was the love of truth for its own sake. This definition carries with it the idea of a love of exactitude— the more exact we are, the nearer we are to the truth. It carries with it a certain independence of authority; because, although an adherence to authoritative propositions taught us by our ancestors, and which we regard as true, may, in a certain sense, be regarded as a love of truth, yet it ought rather to be called a love of these propositions, irrespective of their truth. The lover of truth is ready to reject every previous opinion the moment he sees reason to doubt its exactness. This particular direction of the love of truth will lead its possessor to pursue truth in every direction, and especially to investigate those problems of society where the greatest additions to knowledge may be hoped for.

Scientific method we may define as simply generalized common sense. I believe it was described by Clifford as organized common sense. It differs from the method adopted by the man of business, to decide upon the best method of conducting his affairs, only in being founded on a more refined analysis of the conditions of the problem. Its necessity arises from the fact, that, when men apply their powers of reason and judgment to problems above those of everyday life, they are prone to lose that sobriety of judgment and that grasp upon the conditions of the case which they show in the conduct of their own private affairs. Business offers us an example of the most effectual elimination of the unfit and of "the survival of the fittest." The man who acts upon false theories loses his money, drops out of society, and is no longer a factor in the result. But there is no such method of elimination when the interests of society at large are considered. The ignorant theorizer and speculator can continue writing long after his theories have been proved groundless, and, in any case, the question whether he is right or wrong is only one of opinion.

I ask leave to introduce an illustration of the possibilities of scientific method in the direction alluded to. Looking at the present state of knowledge of the laws of wealth and prosperity of communities, we see a great resemblance to the scientific ideas entertained by mankind at large many centuries ago. There is the same lack of precise ideas, the same countless differences of opinion, the same mass of meaningless speculation, and the same ignorance of how to analyze the problem before us in the two cases. Two or three centuries ago the modern method of investigating nature was illustrated by Galileo, generalized by Bacon, and perfected by Newton and his contemporaries. A few fundamental ideas gained, a vast load of useless rubbish thrown away, and a little knowledge [of] how to go to work acquired, have put a new face upon society. Look at such questions as those of the tariff and currency. It is impossible not to feel the need of some revolution of the same kind which shall lead to certain knowledge of the subject. The enormous difference of opinion which prevails shows that certain knowledge is not reached by the majority, if it is by any. We find no fundamental principles on which there is a general agreement. From what point must we view the problem in order to see our way to its solution?

I reply, from the scientific standpoint. All such political questions as those of the tariff and the currency are, in their nature, scientific questions. They are not matters of sentiment or feeling, which can be decided by popular vote, but questions of fact, as effected by the mutual action and interaction of a complicated series of causes. The only way to get at the truth is to analyze these causes into their component elements, and see in what manner each acts by itself, and how that action is modified by the presence of the others: in other words, we must do what Galileo and Newton did to arrive at the truths of nature. With this object in view, whatever our views of culture, we may let science, scientific method, and the scientific spirit be the fundamental object in every scheme of a liberal education.

What Happened Next

M ore than a century has elapsed since the various essays included in this book were first published. The pace of scientific
discovery has increased enormously since Newcomb's day,
and with it our knowledge of the universe in which we live. Inevitably,
then, there are vast areas of science where little was known in Newcomb's
time but both the broad outlines and, in many cases, fine details have
since been filled in by generations of researchers. And there are other
areas in which what was thought to be established fact turned out to be
incorrect.

That being said, the essays included here were selected—as noted in the
preface—because they still have something useful and interesting to say to
a modern reader. Working through them in order to point out all the ways
in which science has moved on since Newcomb's time would not only be a
tedious exercise, but would also run contrary to the underlying spirit of
the book. After all, I doubt any modern reader has picked up *The
Fairyland of Geometry* expecting an up-to-the-minute overview of any of
the topics he discusses! A hundred years ago, such currency *was* one of
the selling points for Newcomb's science popularizations, but no longer.
What value remains is of a less fleeting nature.

Still, it's interesting at least in passing to note some of the places where
Newcomb got it wrong—and, too, where he got it right. What follows is
nothing more than a series of somewhat casual observations, offered in
the hope that one or two of them may intrigue a reader sufficiently for her
to dig a little deeper into the subject at hand.

As I say in the headnote to Chapter One, "A View of the Universe," it
was generally assumed in Newcomb's time that our own galaxy made up
the entirety of the universe, although some astronomers speculated that at
least some of the nebulae they observed were not nearby clouds of dust or

gas but, instead, much more distant collections of stars—galaxies in their own right. That, in fact, turned out to be the case, though the existence of what for a time were referred to as the "extragalactic nebulae" wasn't firmly established until the 1920s.

In Chapter Two and Chapter Three, Newcomb reviews the accomplishments of nineteenth-century astronomy and looks ahead to the twentieth. Chapter Two is entitled "What the Astronomers Are Doing" and what the astronomers of Newcomb's era were mainly doing was measuring the positions of celestial objects and then cataloguing them. In the days before electronic computers, working out accurate orbital characteristics for the moon or the planets required vast amounts of painstaking mathematical calculation by hand—Newcomb himself spent decades refining his work on the moon's orbit—and cataloguing the positions of the stars was equally tedious before the advent of astronomical photography.

It was only in the late 1830s that astronomers first determined the distance to the nearest stars, and sixty years later accurate parallaxes (and hence distances) had been determined for less than a hundred stars. That number crept slowly upward over the course of the twentieth century but the real breakthrough came with the launch of the Hipparcos satellite in 1989. Data gathered by the satellite allowed astronomers to determine with a high degree of accuracy the distance and position of no less than 120,000 stars—virtually every star within 250 light years of the earth. Somewhat less precise distances are available for more than a million other stars out to a distance of about 1600 light years, and the European Space Agency's Gaia satellite observatory, launched in late 2013, is expected to extend the yardstick to nearly 20,000 light years, providing astronomers with accurate distances for no less than a billion stars.

Back in the closer confines of the solar system, Newcomb in Chapter Three hints at the existence of something that would not be confirmed until the early years of the space age: the "solar wind," a constant outflow of charged particles from the sun. The solar wind's interaction with the earth's own magnetic field and the belts of charged particles that surround our planet give rise to the magnetic storms that perturbed compass needles and so vexed astronomers of Newcomb's day.

In Chapter Three, Newcomb also discusses the problem of the source of the sun's energy. The accepted theory of the day held that the sun produced energy as it slowly contracted. Millions of years ago, the sun had started out as a vast, diffuse cloud of gas wider than the present-day solar system; millions of years from now, it would end its contraction as a cold, dark, solid cinder. Calculations of how quickly the sun had to shrink in order to generate the amount of energy it actually gives off resulted in an upper limit for the solar system's age of about 20 million years. "Here," Newcomb declares, "the geologists step in and tell us that this conclusion is wholly inadmissible" (p. 19): hundreds, even thousands, of millions of years are required to account for the slow unfolding of the geological processes that shaped the face of the earth. Presciently, Newcomb points to the recent discovery of radioactivity as a possible solution to the conundrum. It is indeed the vast energies locked up in the atomic nucleus that have allowed the sun to shine much as it does today for more than five billion years—250 times as long as the estimate worked out by Victorian astronomers. The result is plenty of time for geological processes to do their slow work.

The question of whether life exists beyond the earth was a question that held as much interest in Newcomb's day as in our own. Although we know a lot more now about both life and the universe than we did then, the conclusions Newcomb sets out in the essay "Life in the Universe" are still generally accepted today. Particularly noteworthy to the modern reader is Newcomb's emphasis on the fecundity of life and the wide range of environments it inhabits. That range turned out, of course, to be even wider than Newcomb and his contemporaries suspected: today we know of life forms that thrive near deep-sea hydrothermal vents, deriving energy from the chemicals that are dissolved in the superheated water emerging from beneath the sea floor, as well as microbes and tiny worms that flourish deep underground.

Newcomb's discussion of the possibility of life on Mars is more tempered than the writings of such contemporaries as Percival Lowell, and, as a result, holds up much better a century later. Newcomb was skeptical about the existence of the famous Martian canals that Schiaparelli, Lowell, and some other observers claimed to see, but was willing to concede that, given the presence of water and an atmosphere, albeit a thin one, life might possibly exist on Mars. After half-a-century of exploration by space probes, that conclusion still stands today, though if there is life on Mars it likely exists only beneath the planet's surface where it would be protected from the ultraviolet radiation that, in the absence of an ozone layer like the earth's, reaches the Martian surface in an undiminished, deadly flux. More likely still is the prospect that while life may have existed on Mars near the beginning of the planet's history, it has since perished as environmental conditions deteriorated.

Chapter Five, "Constellation and Star Names," is the only excerpt included in the present volume from Newcomb's 1902 book *The Stars: A Study of the Universe,* which in turn was a compilation of articles that originally appeared in *Popular Science Monthly.* The essay recounts some of the challenges involved in setting accurate and universally accepted boundaries for the constellations. In fact, it wasn't until the 1920s that the International Astronomical Union finally established the number and boundaries of the constellations on the lines still accepted today.

"The Universe as an Organism," the sixth chapter in the collection, is noteworthy mainly for Newcomb's contention that "the trend of recent astronomical and physical science . . . was in the direction of showing the universe to be a connected whole." For instance, the invention of the spectroscope led to the discovery that the stars were made up of the same elements that are found here on earth. And Newcomb was ahead of his time in suggesting that, in addition to light and gravity, "there are other agencies whose exact nature is yet unknown to us, but which do pass from one heavenly body to another." Indeed there are, and much of what we know now about the universe is the product of observations made with such once unknown agencies as gamma rays, X-rays, cosmic rays, neutrinos, and radio waves. Surely Newcomb would have been intrigued by the thought that "dark matter" and "dark energy"—agencies whose origins and nature are even more mysterious to us than were X-rays or radio waves in 1902, when "The Universe as an Organism" was written—play a crucial role in shaping the universe's large-scale structure.

Chapter Seven, "The Coming Total Eclipse of the Sun," has not to my knowledge been reprinted since it first appeared in *McClure's Magazine* in May 1900. Yet it is one of Newcomb's most engaging essays, and one which has dated surprisingly little. Newcomb was a veteran of numerous eclipse expeditions over the course of his career, something which no doubt lends verisimilitude to the description of an eclipse that opens the piece.

One of the most dramatic sights during a total eclipse of the sun is the corona, "an effulgence radiating a saintly glory" (in Newcomb's melodramatic Victorian description) that leaps into visibility when the moon has fully covered the sun's disk. The nature of the corona was still open to debate when Newcomb wrote this article, though it had been established that it was part of the sun, not the moon (as some astronomers had earlier suggested). In fact, the corona is the sun's outer atmosphere; it is heated to a temperature of three million degrees Celsius by electromagnetic mechanisms whose exact nature is still the object of research today. The corona's extremely high temperature explains why it isn't pulled back into the main body of the sun by the latter's immensely strong gravitational field, something which puzzled astronomers of Newcomb's era.

Three other topics raised in the essay deserve note. One is the problem of small deviations in the calculated orbit of Mercury, the innermost planet. Nineteenth-century astronomers thought that the deviations might be the result of the gravitational pull of another, as yet unobserved planet that was even closer to the sun. But the supposed planet was never found (despite several false alarms) and Einstein's theory of general relativity, which refined our understanding of how gravity works, especially under extreme conditions such as those close to a massive body like the sun, successfully accounted for the deviations without any need for a new planet.

Another phantom of nineteenth-century astronomy was the element "coronium," which was thought to have been detected on the sun though it was unknown on earth. As matters turned out, under the extreme temperatures and intense magnetic fields found on the sun, well-known elements such as iron sometimes emit particular wavelengths of light that they do not give off under ordinary conditions. It was these unusual emissions that astronomers had detected using the spectroscope and that had led them to postulate the existence of coronium.

Finally, Newcomb was correct in his assumption that predictions of future eclipses would become ever more accurate. The capacity of modern computers to perform vast numbers of mathematical calculations at lightning speed as well as extremely precise measurements of key figures such as the moon's distance from the earth—now known within a margin of error of only a few centimetres—allow us to accurately predict eclipse times and locations for thousands of years into the future.

The ninth through twelfth chapters of this volume are drawn from *Astronomy for Everybody*. The figures Newcomb cites for such things as the volume or mass of the sun or the major planets are fairly close to the more accurate values we have today. By 1900 the basic plan of the solar system had been sketched in, and until the Second World War only a modest amount of updating was required to keep Newcomb's discussion of the planets reasonably current. With the advent of the Space Age came a tremendous explosion in our knowledge of the solar system. Manned missions to the moon and robotic surveys of all eight major planets and their satellites produced a vast flood of data. A variety of asteroids and comets as well as the sun itself have also been studied in detail by space probes. The result of that half-century of exploration is that for us, the various bodies of the solar system are worlds in their own right, with characteristics and histories as complex as those of our own earth. But for Newcomb and his contemporaries, they were little more than lights in the sky. The sole exceptions to this rule were the moon and Mars.

By the beginning of the twentieth century the side of the moon visible from earth had been mapped to a high degree of detail, and the moon's essential nature had become clear: it was airless, largely or completely waterless, and almost certainly lifeless. And so the first Apollo astronauts found it when they landed in 1969. What astronomers of Newcomb's era did *not* anticipate was the major role played by meteorite impacts in shaping the moon's surface, both on a large and small scale. The moon's craters were formed when meteorites and asteroids of varying sizes collided with the moon. Most of these impacts occurred in the first few hundred million years of the moon's history. Over the course of the billions of years that have elapsed since, a never-ending drizzle of meteoroid and micrometeoroid impacts has chewed up the moon's surface, creating a solidly packed surface layer (the "regolith") of fractured, splintered rock that ranges between two and twenty metres thick. Still, Newcomb was not far from the mark when he summed up the moon as "a world which has no weather and on which nothing ever happens."

When Newcomb wrote *Astronomy for Everybody* the controversy over the Martian canals was at its height. As noted earlier, Newcomb was skeptical about the claims of such canal enthusiasts as Percival Lowell, and in Chapter Twelve he suggests that the canals were optical illusions, caused by the human eye's efforts to try to make sense of irregular details that were at the very edge of visibility. He was right. Although many observers, including the keen-sighted E. E. Barnard, failed to see the canals, belief in their existence lingered in at least some parts of the astronomical community and certainly among the general public well into the 1950s, and it took Mariner 4—the first spacecraft to fly by Mars and return photographs—to put the final nail in their coffin in 1964.

Ironically, the revised edition of *Astronomy for Everybody*, prepared by Robert H. Baker and published in 1932, proved less accurate with regard to the Martian canals than the original book. "There is no longer any doubt as to the existence of the canals of Mars," Baker wrote. "They have been observed by many astronomers, and have been photographed successfully. . . . We accept the canals as natural features of the Martian landscape." —Well, no. Some astronomers claimed to see canals in certain photographs of Mars, but others, looking at the same images, saw nothing, and the canals could not be seen in the reproductions of the photographs printed in newspapers or magazines. Perhaps the same optical illusion that caused disagreement over naked-eye observations of Mars played a part in differing interpretations of these relatively low-resolution early photographs of the planet. In any case, as the photographs improved, the canals vanished.

Baker also took a far more optimistic view of the prospects for life on the red planet than had Newcomb, who, in the original edition of *Astronomy for Everybody,* asked "the reader [to] excuse me from saying anything in this chapter about the possible inhabitants of Mars. He knows just as much of the subject as I do, and that is nothing at all." By 1932 Baker felt comfortable declaring that "There appears to be life on the planet Mars. A few years ago, this statement was commonly regarded as fantastic. Now it is commonly accepted." To be sure, Baker did add the caveat that he was referring "to forms resembling our vegetable life and not to intelligent human life, for there is no evidence to prove the existence of such life on Mars."

The belief that seasonal changes in the planet's color signalled the existence of vegetation persisted until the robotic flybys of the Space Age, and was even bolstered in the late 1950s by spectroscopic observations that appeared to confirm the presence of lichen-like vegetation on Mars. But, as with the canals, observations made at the very limits of an instrument's capability turned out to be wrong. There are no tracts of vegetation on the Martian surface and seasonal color changes are the result of dust being deposited and removed by recurring storms.

The fact that, in the light of modern knowledge of Mars, Baker's 1932 revision of *Astronomy for Everybody* turned out to be less accurate than Newcomb's original 1902 edition is a telling reminder that science does not always progress as smoothly or uniformly as we like to think. Sometimes there are detours; sometimes new research (whether right or wrong) leads us down blind alleys. And sometimes—and this is particularly true of the question of life on Mars—the emotional appeal of a particular conclusion results in evidence that favors that conclusion receiving disproportionate weight, while evidence against it is undervalued or ignored.

Most of Newcomb's astronomical work was highly mathematical in character, and throughout his life he was also interested in various mathematical subjects beyond the boundaries of astronomy, including probability theory. "The Fairyland of Geometry," reprinted as Chapter Thirteen of this collection, is an early popular treatment of non-Euclidean geometry. In it, Newcomb deals with two ideas that were to prove central to twentieth-century cosmology: first, the notion that additional dimensions might exist beyond the familiar three that we know from the physical world around us; and second, the idea that space might be curved rather than flat. He even suggests how a human being might be converted into her mirror image by being flipped over in the fourth dimension: "We should then come back into our natural space, but changed as if we were seen in a mirror. Everything on us would be changed from right to left, even the seams in our clothes, and every hair on our head." This very idea has since served as the basis for several science-fiction stories!

A note on terminology: in the traditional Euclidean geometry of flat planes, it is the fifth postulate (also known as the "parallel postulate") that eventually opened the prospect of non-Euclidean geometries. When Newcomb refers in the essay to the "eleventh axiom of Euclid," he is in fact referring to an edition of Euclid prepared by John Playfair, a Scottish mathematician of the late eighteenth and early nineteenth centuries. Playfair restated Euclid's fifth postulate in a more straightforward way: "Two straight lines, which intersect each other, cannot both be parallel to the same straight line." Although what is now called "Playfair's axiom" and Euclid's fifth are taken as being equivalent by Newcomb, apparently there can be geometries in which that isn't the case. But for our purposes here (and Newcomb's) Euclid's fifth and Euclid's eleventh are one and the same.

The final four essays in the book demonstrate the breadth of Newcomb's interests. He was not an Ivory Tower scientist: his work with the Naval Observatory provided firsthand involvement with politics and government, and he served in administrative roles not only with the Observatory but with a variety of scientific organizations. He also wrote numerous articles on economic issues. Not surprisingly, then, Newcomb was an advocate of applying the scientific method to problems of government and society, and makes the case for doing so in Chapter Fourteen. The question of the proper relationship between science and government is, of course, even more pressing in our day than in his.

Newcomb did not shy away from taking strong positions on issues of public interest. One of these was the feasibility of heavier-than-air flight. Newcomb is perhaps best known today for his failed prediction that heavier-than-air flight was impossible. A little more than two years after

"Is the Airship Coming?" was published, the Wright Brothers proved Newcomb wrong, but Newcomb continued to insist on the impossibility (or at least the impracticality) of the airplane even after one had successfully flown! A later essay, "The Outlook for the Flying Machine," first published in 1903, was reprinted (with some material from "Is the Airship Coming?" added in) in Newcomb's 1906 book *Side-Lights on Astronomy* without any acknowledgement of the Wrights' accomplishment. Admittedly, news—or at least certain kinds of news—spread more slowly then than now, and it was several years before what the Wrights had done became common knowledge. When it finally forced itself on his attention, Newcomb did admit his error in not anticipating how the internal combustion engine, lighter and more powerful than steam engines, would make powered heavier-than-air flight feasible. But in two essays published the year before his death—"The Prospect of Aerial Navigation" in the *North American Review* and "The Problem of Aerial Navigation" in *The Nineteenth Century*—he continued to insist that physical law placed severe limitations on the practicality of heavier-than-air flight, concluding that "the era in which we shall take the flyer as we now take the train belongs to dreamland." This prediction fared little better than the earlier assertion that heavier-than-air flight was completely impossible: the first commercial airline took flight only eight years later.

So what went wrong? It seems too facile to attribute Newcomb's views on flight to lack of imagination. After all, in "Is the Airship Coming?" he happily speculates on the possibility of antigravity, suggesting that when the root causes of gravitation are discovered, "then may men build, not merely airships, but ships which shall fly above the air, and transport their passengers from continent to continent with the speed of the celestial motions" (p. 119).[1]

The problem was that a realistic analysis of the prospects of heavier-than-air flight required a degree of specialized technical knowledge that Newcomb lacked and which, late in life, he was understandably loath to spend the time acquiring. Newcomb draws on various mathematical and scientific principles to argue against heavier-than-air flight—for instance, the square-cube law, which, in brief, holds than an object's volume

---

[1] Arthur C. Clarke, in his *Profiles of the Future* (1962), cites Newcomb as an example of what happens "when *even given all the relevant facts* [emphasis Clarke's] the would-be prophet cannot see that they point to an inescapable conclusion"—what Clarke calls a "failure of nerve" for a prognosticator.

increases more quickly than its surface area. But the points he makes are so simplistic as to have little bearing on the actual technological problem at hand. If a steam engine of a given weight can't generate enough power to make flight practical, that doesn't rule out the possibility that there might be other kinds of engines allowed by physical law that can produce better power-to-mass ratios. Or consider Newcomb's objection that, if a heavier-than-air flying machine experienced mechanical problems, it would crash to earth, killing all its passengers. Given the limitations of early twentieth-century technology, it did seem hard to believe that machines could be developed so reliable that air travel would become the safest mode of mass transportation. But there were no hard-and-fast limits imposed by natural law that *guaranteed* the impossibility of such a prospect. Technical difficulty and technical impossibility are not synonymous.

Newcomb's reservations concerning spiritualism were more solidly grounded than those concerning flight, perhaps because he spent a substantial amount of time researching the former topic, even serving for some years as the president of the American Society for Psychical Research. As a result, "Modern Occultism" (another essay not previously reprinted to my knowledge) is a sensible treatment of the subject that is as sound now as when it was first written. And in "Can We Make It Rain?" (Chapter Seventeen), Newcomb avoids the pitfalls of prediction that befell him in his discussion of flying machines by tightly tying the question at hand to specific scientific principles. The real question addressed in the piece is not whether we can make it rain but whether we can make it rain by firing off cannon and explosives and making loud noises. Newcomb's answer to that very specific question is (quite rightly) no, but he takes pains to point out there might be other scientifically feasible means of creating rain: "For anything we know to the contrary, the agglomeration [of small particles of water suspended in clouds to form raindrops] may be facilitated by smoke in the air. If it be really true that rains have been produced by great battles, we may say with confidence that they were produced by the smoke from the burning powder rising into the clouds and forming nuclei for the agglomeration into drops, and not by the mere explosion."

The final essay, "What Is a Liberal Education?" (Chapter Eighteen), reiterates Newcomb's argument that a solid grounding in the scientific method is the best preparation for life. One suspects that, were he to return today to assess the problems facing humanity a century after his death, he would still maintain that science—"organized common sense," as he called it—remains our best hope and most effective tool for mastering our challenges.

Simon Newcomb: Scientist and Popularizer

S imon Newcomb rose from obscurity and poverty to become one of the best-known and most accomplished scientists of his time. In addition to his research, he was an able administrator and one of the most prolific popularizers of science of his own or, for that matter, any era.

He was born in Nova Scotia to well-educated but impoverished parents. His mother died before he reached adulthood and Newcomb himself suffered some sort of mental breakdown while still a child, finishing his education at home. When he turned 16, he signed on as a doctor's apprentice. The experience was not a happy one: oppressed, taken advantage of, learning nothing, Newcomb eventually fled to the United States, joining his father in Massachusetts and later moving to Maryland to work as a teacher.

From an early age Newcomb had shown mathematical ability far beyond his years, and it was after moving to the United States that he settled on science as his life's work. During breaks from teaching he met Joseph Henry, the director of the Smithsonian Institution. It was Henry who helped Newcomb secure a job at the U.S. Navy's Nautical Almanac Office. In those days, nautical almanacs produced by the Royal Navy, the U.S. Navy, and others were essential tools of navigation, providing exact positions of various celestial objects for a multitude of times and positions. This enabled seafarers to compare their own sightings of celestial bodies with the almanac's figures and so determine their position on the ocean with a high degree of accuracy. During this period, Newcomb also earned a bachelor of science degree from Harvard's Lawrence Scientific School.

Work on the orbital dynamics of the asteroids first earned Newcomb widespread recognition in the astronomical community: he carried out calculations showing that the observed orbits of the asteroids were incompatible with the widely held theory that they were the remnants of one original planet that had somehow exploded or broken apart. (Despite Newcomb's work, the "exploding planet" theory of the origin of the asteroids remained a fixture of textbooks and popular accounts for at least another century.) Newcomb's work on orbital dynamics may also link him to a famous fictional character. At least one historian of science has advanced Newcomb as the inspiration for Doctor Moriarity, the master criminal and Sherlock Holmes's famous arch-nemesis, whose (fictional) book *The Dynamics of an Asteroid* was described by Holmes as "a book which ascends to such rarefied heights of pure mathematics that it is said that there was no man in the scientific press capable of criticizing it."

It would seem, however, that Newcomb's resemblance to Moriarity began and ended with mathematical acuity. True, Newcomb has been characterized by modern biographers as "outspoken and often openly partisan," an "astronomer with an attitude." And over the course of his long career he never shied away from controversy or debate. But in person he was well-liked and sociable, described by contemporaries such as the Progressive politician and author Frederic C. Howe as "a big, lusty, joyous man."[1] Nor did Newcomb share Moriarity's passion for secrecy. Newcomb's work in government and academe as well as his prolific writing career made him instead the best-known scientist of the age.

In 1861, Newcomb became professor of mathematics at the Naval Observatory. Later appointments included a stint as superintendent of the same Nautical Almanac Office where he had begun his career, a professorial appointment at Johns Hopkins, and the presidencies of several academic and scientific organizations, including the American Association for the Advancement of Science. He retired from the Observatory in 1897 but remained active until his death in 1909; indeed, the final decade of his life was perhaps his most prolific as a writer. During the years leading up to his death he also continued to work intensively on the scientific problem that had engaged him throughout his career: more accurately calculating the motions of the planets and, especially, those of the moon. The results derived from his calculations remained standard until the advent of the Space Age.

---

[1] The quotes are from Albert E. Moyer and Alfred T. Kamajian, "Simon Newcomb: Astronomer with an Attitude," *Scientific American* 279 (October 1998).

Astronomy boasts a long tradition of prominent practitioners of the science who were also first-rate popularizers of the discipline, but Newcomb's combination of first-rate research with a wide popular presence was probably not rivalled until Carl Sagan's rise to fame in the 1970s. The bibliography published by the National Academy of Sciences after Newcomb's death lists not only dozens of scientific papers but even greater numbers of popular articles, textbooks, book reviews, and encyclopedia articles. His articles on science were regular fixtures in such magazines as *Harper's*, *McClure's*, and *Popular Science Monthly*, and his 1902 book *Astronomy for Everybody*, based on previously published magazine pieces, went on to sell more than 50,000 copies in various editions in the United States and Great Britain. (It was revised in 1932 by Robert H. Baker and new printings were still appearing at least as late as 1943.)

Nor did Newcomb restrict his choice of subjects to astronomy and mathematics. He held strong views on economics and published nearly four dozen books and articles in the field, including *The ABC of Finance; or, The Money Question Familiarly Explained to Every-day People in Nine Short and Easy Lessons* (1877), and a textbook, *Principles of Political Economy* (1886), which would later win praise from no less prominent an economist than John Maynard Keynes as an original contribution to the discipline.

A selection of titles chosen at random from the National Academy of Sciences bibliography illustrates how broad Newcomb's interests tended to be: science, mathematics, and economics aside, it includes everything from "Sentimentalism in Politics" (*The Nation*, 1879) to "The Plan of the Bosses at Chicago" (an anonymous editorial also published in *The Nation*, 1880) to a letter on the Swiss watch industry in *Science* (1883). Other pieces are entitled "The Possibilities of Invention" (New York *Independent*, 1899), "Science and Government" (*North American Review*, 1900), "The Functions of the Senate" (*The Nation*, 1903), and "University Athletics" (*North American Review*, 1907). Newcomb also penned three science-fiction stories, including a full-length novel, *His Wisdom, The Defender* (1900), which was reviewed favorably in *The Nation*. Late in life he also published his autobiography, *The Reminiscences of an Astronomer* (1903).

Popular science writing has its fashions as does everything else. Newcomb's prose seems old-fashioned by modern standards, but no doubt twenty-second century readers will say the same of books and articles published in our own time. The hallmarks of successful science writing, however, don't change, and in the reviews of Newcomb's works the same adjectives appear—"lucid," "clear," "graceful," "readable"—that one expects to find in a favorable review of like works published today. When he died in 1909, he left behind not only a substantial body of research, but a lifetime of accomplishment as a popularizer and interpreter of science that would not be soon surpassed.

FURTHER READING AND SOURCES

The sole full-length biography of Newcomb is by Bill Carter and Merri Sue Carter: *Simon Newcomb: America's Unofficial Astronomer Royal* (Mantanzas Publishing, 2006). He has also been the subject of at least two relatively recent magazine articles: "Simon Newcomb's Journey" by Jean-Louis Trudel (*The Beaver,* December 2003—there is also a sidebar, "Was Newcomb the Original Moriarty?"), and "Simon Newcomb: Astronomer with an Attitude" by Albert E. Moyer and Alfred T. Kamajian (*Scientific American,* October 1998). All of these are readable, engaging introductions to this multi-faceted man. Allowing for the conventions of Victorian biography and autobiography, Newcomb's own memoir, *The Reminiscences of an Astronomer* (1903), is also of interest.

Helpful in preparing this book was W.W. Campbell's biographical memoir of Newcomb, published by the National Academy of Sciences in 1916, and the comprehensive bibliography prepared by Raymond Clare Archibald, published by the Academy in 1924. Various standard sources, including the entries on Newcomb in the *Dictionary of Scientific Biography* and the *Dictionary of Canadian Biography Online* were also consulted.

Eugene Garfield offers an intriguing take on Newcomb's views on heavier-than-air flight in his "Negative Science and 'The Outlook for the Flying Machine,'" included in *Essays of an Information Scientist,* Volume 3 (ISI Press, 1980). Newcomb and other aero-skeptics are discussed in some detail in Arthur C. Clarke's *Profiles of the Future: An Enquiry into the Limits of the Possible* (Harper, 1962), which also introduces "Clarke's Law," viz.: "When a distinguished but elderly scientist states that something is possible, he is almost certainly right. When he states that something is impossible, he is very probably wrong."

Editor's Notes

The purpose of this book is to introduce Simon Newcomb's writings on science to a new generation—indeed, a new century!— of readers, in the belief that while science marches on, there are elements of the best science writing of the past that remain of interest and value to today's readers and deserve to be rediscovered.

So I should be clear on what *The Fairyland of Geometry* is not. It is *not* meant to be a scholarly edition, complete with apparatus and notes. Nor have all the peculiarities of early twentieth-century spelling, grammar, and usage been slavishly preserved. (Peculiarities to us, anyway: no doubt some of the conventions of our own day would seem just as odd to Newcomb, were he able to see them!) Where appropriate, spelling has been modernized, and although care has been taken to preserve Newcomb's unique voice, in those very few cases where the sense of a word or phrase has changed drastically since his day, the necessary change has been silently made.

Other than these minor stylistic edits, the articles, essays, and book chapters included in *The Fairyland of Geometry* appear for the most part just as they were first published. In a few cases, material so topical as to be of no interest to the modern reader has been excised (for instance, a discussion of future solar eclipses—future, that is, to Newcomb's readers; long past and forgotten to us). In the chapters drawn from *Astronomy for Everybody*, internal references to other sections of the book not reprinted here have also been deleted. In each case, an ellipsis (". . .") indicates where material has been cut. In no case save one discussed below has the basic sense of the prose been altered by the cuts.

The one exception involves a section in Chapter 4, "Life in the Universe," in which Newcomb discusses life's ability to withstand extremes of hot and cold. Pondering the perceived backwardness of peoples living in the earth's equatorial regions (a widely held view during Newcomb's time; think no further than Kipling's famous poem "The White Man's Burden"), Newcomb wonders whether their lack of progress should be attributed to the enervating effects of tropical climates or, instead, to innate inferiority.

He settles on the latter reason. Here is the paragraph in question, from p. 121 of *Side-lights on Astronomy*:

> It has often been said that this [i.e., that warmer climates create a greater blossoming of life] does not apply to the human race, that men lack vigor in the tropics. But human vigor depends on so many conditions, hereditary and otherwise, that we cannot regard the inferior development of humanity in the tropics as due solely to temperature. Physically considered, no men attain a better development than many tribes who inherit the warmer regions of the globe. The inferiority of the inhabitants of these regions in intellectual power is more likely the result of race heredity than of temperature.

I decided to excise the entire paragraph from the body of the essay. This was done not out of any desire to present Newcomb as something other than what he was: he was a man of his time, and that time's attitudes about such things as race and gender were far different than those prevalent today. If anything, it's surprising how little one finds in Newcomb's writings that jars modern sensibilities, and there is some evidence that his views were relatively enlightened: he is credited in playing a part in gaining the admission of the first African-American student to Johns Hopkins. But regardless of Newcomb's views on race, in this particular instance the observation being made did nothing to advance the essay's underlying argument, and merely distracts a present-day reader from considering what are otherwise surprisingly modern views of life's place in the universe.

As far as illustrations are concerned, I have included either reproductions or re-drawings of those few that Newcomb directly refers to in his text, and have added a couple of others of interest—for instance, one of the pioneering eclipse photographs that ended up being taken during the eclipse that Newcomb previewed for his readers in "The Coming Total Eclipse of the Sun."

Finally, I toyed with the idea of extensively annotating the text, in the process providing additional biographical detail on persons that Newcomb mentions in passing, commentary on subsequent discoveries and theories, and so on, but in the end—apart comments made in the book's end matter—chose not to. Instead, I have let Newcomb, as much as possible, speak for himself. I think the great polymath and popularizer still has something worthwhile to say.

About the Editor

David Stover is the co-author of two books about the Nobel Prize-winning neuroscientist Roger W. Sperry: *Beyond a World Divided: Human Values in the Brain-Mind Science of Roger Sperry* (1991) and *A Mind for Tomorrow: Facts, Values, and the Future* (2000), both with Erika Erdmann, as well as numerous newspaper and magazine articles, many of them dealing with various aspects of science. He has spent most of his career in book publishing in sales, marketing, editorial, and general management roles with Prentice Hall, McGraw Hill, Allyn & Bacon, Pearson, and the Canadian branch of Oxford University Press, where he was appointed president in 2006.

Stover holds a master's degree in journalism from the University of Southern California as well as an M.A. in history from the University of Western Ontario, where he carried out research on the popularization of science in eighteenth-century England.

Milestones in Science and Discovery

Hundreds if not thousands of compellingly readable works of science exposition and popularization have been published during the past 150 years. Science, of course, marches on, but that does not mean that these books, many of them long out-of-print and almost forgotten, have nothing to say to modern readers. *Milestones in Science and Discovery* seeks to bring these classic works—many of them written by giants both of research and of science popularization—back into circulation in handsome new print and ebook editions designed with the needs of today's readers in mind.

After all, science, like all fields of human endeavour, is cumulative; and in many fields the basic structure was already well-established by the mid-twentieth century. The cutting edge of modern theoretical physics may lie in the exotic realm of string theory, but the foundational worldviews that underlie much of the technology of the modern world—quantum theory and relativity—were first formulated a century ago and well-established by the 1950s. And, of course, many of the key ideas underpinning science as a whole are older yet: Newton's laws of motion and theory of universal gravitation date from the seventeenth century, and Darwin's *Origins of Species* first saw print in 1859. Even the discovery of the basic mechanism of heredity—the famous double helix of DNA—is now sixty years in the past.

Yet countless important works of popular science—books as lively, relevant, and readable now as when first published—today are out of print, discarded from libraries, and unavailable except as musty used copies or murky scans of the original book blocks. The loss is not only to science but literature as well, for many of these authors were master stylists with much to say about the human condition as well as science itself.

All this being said, science *does* continue to progress. And so it is not enough to merely reprint a classic work of science exposition without comment. The work must be put in context, with a sense given of the field's progress in the years since the book was published.

Such is the thinking behind *Milestones in Science and Discovery*, its mission to re-publish renowned works of science popularization in attractive new editions, complete with new material that places these seminal books in a contemporary context.